Photographic and Descriptive Musculoskeletal

Atlas of Orangutans

With notes on the attachments, variations, innervation, function and synonymy and weight of the muscles

Photographic and Descriptive Musculoskeletal Atlas of Orangutans

With notes on the attachments, variations, innervation, function and synonymy and weight of the muscles

- Rui Diogo
- Juan F. Pastor
- Eva M. Ferrero
- Mercedes Barbosa
- Julia Arias-Martorell

- Josep M. Potau
- Félix J. de Paz
- Gaëlle Bello
- M. Ashraf Aziz
- Bernard A. Wood

CRC Press
Taylor & Francis Group
Boca Raton London New York

CRC Press is an imprint of the
Taylor & Francis Group, an **informa** business

A SCIENCE PUBLISHERS BOOK

CRC Press
Taylor & Francis Group
6000 Broken Sound Parkway NW, Suite 300
Boca Raton, FL 33487-2742

First issued in paperback 2019

ISBN-13: 978-1-4665-9727-3 (hbk)
ISBN-13: 978-0-367-38001-4 (pbk)

Visit the Taylor & Francis Web site at
http://www.taylorandfrancis.com

and the CRC Press Web site at
http://www.crcpress.com

Preface

Orangutans, together with chimpanzees and gorillas, are our closest living relatives. This photographic and descriptive musculoskeletal atlas of *Pongo* is the last book of a series of atlases of all apes designed to help provide the comparative, phylogenetic, and evolutionary context for understanding the evolutionary history of the gross anatomy of modern humans and our closest relatives. The atlas includes the results of an extensive review of the literature about the musculature of orangutans, a comprehensive review of muscle variants among individual members of the genus *Pongo*, and a list of the synonyms used in the literature to refer to the muscles of these members of this genus. We hope this atlas will be of interest to students, teachers and researchers studying primatology, comparative anatomy, functional morphology, zoology, and physical anthropology, as well as to clinicians and researchers who are interested in understanding the origin, evolution, and homology of the musculoskeletal system of modern humans as well as the comparative context of common variants on the musculature of modern humans.

April 6th, 2013

Acknowledgements

We gratefully acknowledge support and funding from all of the institutions and funding bodies that made this project possible; we especially acknowledge support for RD's research via a GW Presidential Fellowship and a Howard University College of Medicine start-up package. We particularly acknowledge B. Richmond (George Washington University, US) for allowing us to dissect the GWU PP1 specimen, the Zoo-Aquarium of Madrid for providing the VU PP1 specimen, the Bioparc Fuengirola of Malaga for providing the VU PP2 specimen, and the Zoo de Santillana (Cantabria) (all from Spain) for providing the VU PA1 specimen.

Contents

CHAPTER 1

Introduction and Aims

Orangutans, together with chimpanzees and gorillas, are our closest living relatives. This photographic and descriptive musculoskeletal atlas of *Pongo* follows the organization used in the photographic atlases of *Gorilla*, *Hylobates*, and *Pan* published by Diogo et al. in 2010, 2012 and 2013, respectively. This is thus the last book of a series designed to help provide the comparative, phylogenetic, and evolutionary context for understanding the evolutionary history of the gross anatomy of modern humans and our closest relatives.

We dissected and took high-quality photographs of 5 specimens of the two extant orangutan species (*P. pygmaeus* and *P. abelii*), including neonates and adults and both males and females (see Methodology and Material below) and for two of these specimens (VU PP1, VU PP2) we were able to record the wet weight of many of the muscles. Where there are differences between the myology (e.g., the presence/absence of a muscle or a muscle bundle, its attachments and/or its innervation) of the specimen PP1 and any of the other specimens dissected by us, we provide detailed comparative notes, and use photographs to document the differences.

The atlas also includes the results of an extensive review of the literature about the musculature of orangutans, a comprehensive review of muscle variants among individual members of the genus *Pongo*, and a list of the synonyms used in the literature to refer to the muscles of these members of this genus. The data previously obtained from our dissections of numerous primates and other mammalian and non-mammalian vertebrates (e.g., Diogo 2004a,b, 2007, 2008, 2009, Diogo & Abdala 2007, 2010, Diogo et al. 2008, 2009a,b, 2010, 2012, Diogo & Wood 2011, 2012) were used to test hypotheses about the homologies among the muscles of orangutans, other apes and modern humans, and other taxa.

We hope this atlas will be of interest to students, teachers and researchers studying primatology, comparative anatomy, functional morphology, zoology, and physical anthropology, as well as to clinicians and researchers who are interested in understanding the origin, evolution, and homology of the musculoskeletal system of modern humans as well as the comparative context of common variants on the musculature of modern humans.

CHAPTER 2

Methodology and Material

The five orangutans examined in this study were made available by the following institutions: George Washington University (*P. pygmaeus*, GWU PP1, adult male, formalin embalmed), Howard University (*P. pygmaeus*, HU PP1, neonate male, formalin embalmed), and Valladolid University (*P. pygmaeus*, VU PP1, adult female, fresh, provided by the Zoo-Aquarium of Madrid; *P. pygmaeus*, VU PP2, adult female, fresh, provided by the Bioparc Fuengirola of Malaga; *P. abelii*, VU PA1, adult female, skeleton, provided by the Zoo de Santillana of Cantabria). We took photographs of the musculoskeletal system of all of the specimens we dissected; the muscle weights listed in this atlas are from the VU PP1 and VU PP2 specimens (total body weight = 72 and 61 kilograms, respectively), which were in particularly good condition. The photographs of the osteological structures shown in this atlas are from the VU PA1 and VU PP2 specimens. In the text below, when the data are available, we provide for each muscle: 1) its weight in the VU PP1 and VU PP2 specimens. The total mass of all the striated muscles is given in parentheses immediately following the name of the muscle. In the case of paired muscles, the muscles of the left and right sides are referred to as LSB and RSB, respectively. When the muscle is part of a symmetrical structure (e.g., stylohyoideus), the weight given is that of the muscle of one side of the body; when the muscle is unpaired (e.g., diaphragm), the weight given is that of the part of the muscle that comes from one side only; 2) the most common attachments and innervation of the muscle within the orangutan clade, based on our dissections and on our literature review; 3) the function of the muscle (e.g., based on electromyographic—EMG—studies); 4) comparative notes, especially when where there are differences (e.g., regarding the presence/absence of the muscle, or of its bundles, its attachments, and/or its innervation) between the configuration usually found in orangutans and the configuration found in a specimen dissected by us (in these cases we often provide photographs to illustrate the differences) or by others; and 5) a list of the synonyms that have been used by other authors to designate that muscle.

Apart from the orangutan specimens mentioned above, we have dissected numerous specimens from most vertebrate groups, including bony fish, amphibians, reptiles, monotremes, rodents, colugos, tree-shrews, and numerous primates, including modern humans (a complete list of these specimens and the terminology used to describe them is given in Diogo & Abdala 2010 and Diogo & Wood 2011, 2012). This broad comparative context proved to be crucial for generating hypotheses about the homologies among the muscular structures of orangutans, modern humans and other primate and non-primate vertebrates, and it also informed the nomenclature we proposed by Diogo et al. (2008, 2009a,b), Diogo & Abdala (2010) and Diogo & Wood (2011, 2012). This nomenclature is based on that employed in modern human anatomy (e.g., Terminologia Anatomica 1998), but it also takes into account the names used by researchers who have focused on non-human mammals (e.g., Saban 1968, Jouffroy 1971). In the majority of the figures we use Latin names for the soft tissues and anglicized names for the bones. In the figures that mainly illustrate osteological structures we use Latin names, but to avoid redundancy when these names are similar to the anglicized version (e.g., processus mastoideus = mastoid process) we do not provide the latter; in those cases in which they are substantially different (e.g., incisura mandibulae and mandibular notch) we provide both the Latin names and the anglicized version.

When we describe the position, attachments and orientation of the muscles and we use the terms anterior, posterior, dorsal and ventral in the sense in which those terms are applied to pronograde tetrapods (e.g., the sternohyoideus mainly runs from the sternum, posteriorly, to the hyoid bone, anteriorly, and passes mainly ventrally to the larynx, which is, in turn, ventral to the esophagus; the flexors of the forearm are mainly situated on the ventral side of the forearm). However, the nomenclature used in Terminologia Anatomica (1998) was defined on the basis of an upright posture and although most primates are not bipeds, nearly all of the osteological names and most of the myological ones used by other authors (and by us) to designate the structures of non-human primates, including orangutans, follow the Terminologia Anatomica nomenclature. Although this is potentially confusing we judged it to be preferable to refer to the topology of the musculoskeletal structures of non-human primates in this way because in the vast majority of primates the 'superior angle of the scapula' is actually mainly anterior, and not superior, to the 'inferior angle of the scapula', and the 'cricoarytenoideus posterior' actually lies more on the dorsal, and not on the posterior, surface of the larynx. Moreover, we think that, by keeping in mind that the actual names (both in Latin and in English) of all the osteological structures and of most the myological structures mentioned in this atlas refer to a biped posture while the actual descriptions provided here regarding the topology of these structures refer to a pronograde posture, most readers will have no difficulty in interpreting and understanding the information provided in this book.

The muscles listed below are those that are usually present in adult orangutans; muscles that are only occasionally present in adult orangutans are discussed in other parts of the atlas. In our written descriptions, we follow Edgeworth (1935), Diogo & Abdala

(2010) and Diogo & Wood (2011, 2012) and divide the head and neck muscles in five main subgroups: 1) mandibular, muscles that are generally innervated by the fifth cranial nerve (CN5) and include the masticatory muscles, among others; 2) hyoid, muscles that are usually innervated by CN7 and include the facial muscles, among others; 3) branchial, muscles that are usually innervated by CNC9 and CN10, and include most laryngeal and pharyngeal muscles, among others; 4) hypobranchial, muscles that include all the infrahyoid and tongue muscles, and the geniohyoideus. According to Edgeworth (1935) the hypobranchial muscles were developed primarily from the anterior myotomes of the body and then migrated into the head. Although they retain a main innervation from spinal nerves, they may also be innervated by CN11 and CN12, but they usually do not receive any branches from CN5, CN6, CN7, CN8, CN9 and CN10; 5) extra-ocular, muscles that are usually innervated by nerves CN3, CN4 and/or CN6 in vertebrates. The head, neck, pectoral and upper limb muscles are listed following the order used by Diogo et al. (2008, 2009a), Diogo & Abdala (2010) and Diogo & Wood (2011, 2012), while the pelvic and lower limb muscles, as well as the other muscles of the body, are listed following the order used by Gibbs (1999). It should be emphasized that the literature review undertaken by this latter author provided a crucial basis and contribution for our own literature review.

CHAPTER 3

Head and Neck Musculature

3.1 Mandibular musculature

Mylohyoideus (VU PP1 LSB = 20.5g; Fig. 9)
- Usual attachments: From the mylohyoid line of the mandible to the hyoid bone, posteriorly, and to the ventral midline, anteriorly.
- Usual innervation: Mylohyoid nerve of mandibular division of CN5 (Wall et al. 1994).
- Function: Stimulation of the mylohyoideus in *Pongo* elicited slight hyoid elevation and tongue protrusion (Wall et al. 1994).
- Notes: There is usually no distinct median raphe of the mylohyoideus in orangutans, according to Sonntag (1924a) and Brown & Ward (1988). The **intermandibularis anterior** (present in some mammals) is not present as a distinct muscle in orangutans. Winkler (1991) suggested that in *Pongo* the **digastricus anterior** is fused with the mylohyoideus, as found occasionally in *Homo*. However, most authors agree that the digastricus anterior is usually missing in orangutans (e.g., Owen 1830-1831, Chapman 1880, Sonntag 1924a, Cachel 1984, Richmond 1993, Wall et al. 1994, our dissections), except in, e.g., one specimen dissected by Parsons (1898a), in which the anterior digastrics were in contact at the midline (as it is usually the case in *Hylobates* and *Pan* but not in *Homo* and *Gorilla*). Cachel (1984) stated that the absence of the anterior digastric may be due to the fact that the large size and weight of *Pongo* mandible may be sufficient to achieve full mandibular depression without this muscle. Brown & Ward (1988) hypothesized that the hyolaryngeal specializations in *Pongo* are responsible for the absence of this muscle, while Winkler (1991) suggested that diet specializations may have also played a role in the absence of the digastricus anterior as a distinct muscle. Wall et al. (1994) stated that the absence of the anterior digastric in *Pongo* results in a decoupling between the hyoid movements and mandibular depression (the digastricus anterior and digastricus posterior in, e.g., gibbons, are usually more related to hyoid movements and mandibular depression, respectively), and that

during unilateral activity the posterior digastric may work with the lateral pterygoid and potentially contribute to substantial transverse movements of the mandible.
- Synonymy: Intermandibularis (Edgeworth 1935).

Tensor tympani
- Usual attachments: From the auditory tube and adjacent regions of the neurocranium to the manubrium of the malleus.
- Usual innervation: Data are not available.

Tensor veli palatini
- Usual attachments: From the Eustachian tube and the adjacent regions of the cranium (usually the scaphoid fossa: Dean 1985) to the pterygoid hamulus and soft palate.
- Usual innervation: Data are not available.
- Notes: In modern human infants and in adult apes, including orangutans, the palate lies much closer to the roof of the nasopharynx than it usually does in adult modern humans, so in the former the levator veli palatini and tensor veli palatini do not run so markedly downwards to reach the palate as they do in the latter. The **pterygotympanicus** (present, e.g., as an anomaly in modern humans) is usually missing in orangutans.
- Synonymy: Tensor palatini (Sonntag 1924a).

Masseter (Figs. 2, 5–6, 9–11, 26)
- Usual attachments: Mainly from the zygomatic arch; the pars superficialis (Fig. 5) inserts mainly onto the lower edge of the base of the mandible, while the pars profunda (Fig. 5) inserts mainly onto the ascending ramus and the coronoid process of the mandible.
- Usual innervation: Branch of the mandibular division of CN5 (Winkler 1991).
- Function: Stimulation of the superficial head of the masseter in *Pongo* elicited mandibular elevation and protraction (Wall et al. 1994).
- Notes: As in gorillas and hylobatids, but contrary to chimpanzees, in orangutans there is often no strong fascia/aponeurosis between the superficial and deep heads of the masseter (e.g., Sonntag 1924a, Boyer 1935a). The **zygomatico-mandibularis** is present in *Pongo* according to Saban (1968) and seemingly according to Boyer's (1935a) descriptions, but this structure was not described by Sonntag (1924a) and was not found in our dissections.

Temporalis (Figs. 5, 26)
- Usual attachments: From the fossa temporalis and the temporalis fascia to the coronoid process and ramus of the mandible.
- Usual innervation: Branch of the mandibular division of CN5 (Winkler 1991).
- Notes: In orangutans the temporalis is usually not divided into a distinct pars superficialis and a distinct pars profunda, and there is also no distinct pars suprazygomatica (Boyer 1935a, our dissections).

Pterygoideus lateralis
- Usual attachments: From the lateral pterygoid plate and the adjacent regions of the cranium (e.g., great wing of the sphenoid bone: e.g., Gibbs 1999) to the capsule of the temporomandibular joint and the neck of the mandibular condyle.
- Usual innervation: Branch of the mandibular division of CN5 (Winkler 1991).
- Function: Stimulation of the inferior head of the pterygoideus lateralis in *Pongo* elicited pronounced mandibular depression (Wall et al. 1994).
- Notes: The inferior and superior heads of the pterygoideus lateralis are usually differentiated in orangutans (Sonntag 1924a, Boyer 1935a, Wall et al. 1994).
- Synonymy: Pterygoideus externus (Sonntag 1924a, Boyer 1935a).

Pterygoideus medialis
- Usual attachments: Mainly from the medial surface of the lateral pterygoid plate of the sphenoid bone and at least sometimes also from adjacent regions of the skull such as and the pyramidal process of the palatine bone, the tuberosity of the maxilla and/or the pterygomandibular raphe (Boyer 1935a, Gibbs 1999), to the medial side of the mandible.
- Usual innervation: Branch of the mandibular division of CN5 (Winkler 1991).
- Synonymy: Pterygoideus internus (Sonntag 1924a, Boyer 1935a).

3.2 Hyoid musculature

Stylohyoideus (Fig. 5)
- Usual attachments: Mainly from the styloid process to the hyoid bone.
- Usual innervation: Data are not available.
- Notes: Contrary to humans and other apes, there is usually no piercing of the stylohyoideus by the intermediate digastric tendon in orangutans (Sonntag 1924a). To our knowledge the **jugulohyoideus** (usually present in strepsirrhines and sometimes present in *Tarsius*) has never been reported in orangutans. However, a muscle **stylolaryngeus** is sometimes present in these apes, which is often a thin structure running from the styloid process to the laryngeal sac (Fick 1895a,b, Sonntag 1924a,b, Falk & Nicholls 1992, Falk 1993).

Digastricus posterior (VU PP1 LSB = 18.5 g; Figs. 5, 8–9, 26)
- Usual attachments: From the mastoid region and adjacent regions of the skull (e.g., sometimes from the occipital bone: Sonntag 1924a) to the back of the mandible.
- Usual innervation: Branch of CN7 (Winkler 1991).
- Function: According to Winkler (1991) and Richmond (1993) in *Pongo* the posterior digastric seemingly acts only in depressing the mandible; however, stimulation of this muscle in *Pongo* by Wall et al. (1994) elicited mandibular depression and slight retraction of the mandible.

- Notes: Contrary to all other extant primates (in which the anterior portion of the digastricus posterior is usually connected to the posterior portion of the digastricus anterior) in *Pongo* the anterior portion of the digastricus posterior is usually directly attached onto the back of the mandible (e.g., Owen 1830-1831, Chapman 1880, Sonntag 1924a, Cachel 1984, Winkler 1991, Richmond 1993, Wall et al. 1994, our dissections). As explained above, in an orangutan specimen described by Parsons (1898a) there was a distinct digastricus anterior, and the anterior portion of the digastricus posterior was connected to this muscle, as is the case in other primates, and not to the back of the mandible, as is the almost always the case in *Pongo*. Another rare configuration for orangutans was reported by Lightoller (1939): Fig. 37 of his plate 8, shows a specimen in which the digastricus posterior inserts mainly onto the angle of the mandible, as is usually the case in the members of this genus, but also sends a very thin musculotendinous slip to the hyoid bone, near the insertion of the stylohyoideus.

Stapedius
- Usual attachments: Probably inserts onto the stapes, but the information provided in the literature is sparse.
- Usual innervation: Data are not available.

Platysma myoides (VU PP1 LSB platysma myoides + platysma cervicale = 90.0 g; Figs. 1–3, 19, 25)
- Usual attachments: Main from the pectoral region and the neck to the modiolus and adjacent regions of the mouth.
- Usual innervation: Data not available.
- Function: According to Winkler (1989) in *Pongo* facial muscles such as the 'platysma', the zygomaticus, the frontalis, the orbicularis oculi, and even the facial auricular muscles, act to support the cheek fat pads, so their size is correlated with the presence/size of these pads, e.g., they are much more developed in adults with well-developed pads.
- Notes: The **sphincter colli superficialis** and **sphincter colli profundus** are usually not present as distinct muscles in orangutans. Influential authors such as Owen (1830-1831) and Sonntag (1924a,b) used the name 'platysma myoides' to describe the platysma complex of Asian apes such as orangutans, and this nomenclature has been followed by various researchers, including Seiler (1976), and was thus also followed in Diogo et al.'s (2009b) review of the mammalian facial muscles. However, Owen (1830-1831) stated that the 'platysma myoides' of orangutans incorporates the platysma myoides of modern humans plus the platysma cervicale of other mammals, and Fig. 1 of Sonntag (1924a) corroborates this statement. The statement was also corroborated by numerous other authors, including Deniker & Boulart (1885), Sullivan & Osgood (1925), Lightoller (1928a, (1940a), Huber (1930b, 1931),

Edgeworth (1935) and Winkler (1989), who reported that juvenile and adult orangutans have a well developed platysma cervicale ('notoplatysma'). Our recent dissections of numerous primates and our comparisons with the data provided in the literature corroborate the statements of all of these latter authors (Figs. 1–3, 25). That is, juvenile and adult orangutans and hylobatids usually have a well-developed platysma cervicale similar to the muscle that is found in most other primates and that is usually markedly reduced, or even absent, in *Pan* (including the neonates dissected by us), *Gorilla* and *Homo*. In fact, juveniles and adult members of these three later genera usually do not have a well-developed platysma cervicale, often having instead a small muscle 'transversus nuchae', which seems to be a vestige of the platysma cervicale (e.g., Loth 1931, Diogo & Wood 2011, 2012).

- Synonymy: Subcutaneous colli or tracheloplatysma (Sullivan & Osgood 1925); mainly corresponds to tracheloplatysma *sensu* Lightoller (1928a); part of platysma myoides (Seiler 1976); part of platysma (Winkler 1989).

Platysma cervicale (VU PP1 LSB platysma myoides + platysma cervicale = 90.0 g; Figs. 1–3, 25)
- Usual attachments: Main from the nuchal region to the modiolus and adjacent regions of the mouth.
- Usual innervation: Branch of CN7 (Gibbs 1999).
- Notes: See platysma myoides above.
- Synonymy: Part of notoplatysma (Lightoller 1928a); part of platysma cervicale (Seiler 1976); part of platysma (Winkler 1989).

Occipitalis (Figs. 1, 4)
- Usual attachments: Mainly from the occipital region to the galea aponeurotica.
- Usual innervation: Branch of CN7 (Gibbs 1999).
- Notes: In hylobatids and non-hominoid primates the occipitalis is usually differentiated into a main body (or 'occipitalis proprius') and a 'cervico-auriculo-occipitalis' (*sensu* Lightoller 1925, 1939, 1940a,b). The latter is a lateral/superficial bundle of the occipitalis that often runs anterolaterally from the occipital region to the posterior portion of the ear and that sometimes covers part of the auricularis posterior in lateral view. The 'cervico-auriculo-occipitalis' of primates such as *Macaca* was designated as the 'deep layer of the occipitalis' by Huber (1930b, 1931), but it is not homologous to the 'pars profunda' found in *Pongo* by Sullivan & Osgood (1925) and by us (Fig. 4). Our dissections and the literature reviewed by us (for detailed accounts on this issue see, e.g., Lightoller 1928a, Seiler 1976) confirm that the 'cervico-auriculo-occipitalis' is usually not present as a distinct structure in *Pongo*. The 'pars profunda' described by Sullivan & Osgood (1925) is mainly a transversal structure, like "a continuation of the sternocleidomastoideus". According to these authors

Sullivan & Osgood this 'pars profunda' is "a small structure of doubtful homology which lies deep to the auriculo-occipital muscle; it has an origin for about 10 mm along the middle of the superior nuchal line; from here the fibers pass upward, fleshy for about 20 mm, at right angles to the fibers of the auriculo-occipital muscle; it continues the direction of the sterno-cleido-mastoid muscle; it ends in the deeper layer of fascia of the cranial vault". They describe a 'pars nuchalis' of the occipitalis, which might correspond to the 'cervico-auriculo-occipitalis' *sensu* the present study; this 'pars nuchalis' has two small bundles, each 2 or 3 mm wide, originating from the external occipital protuberance and then becoming mainly tendinous, blending with the 'pars superficialis' (i.e., the main body) of the occipitalis. However there is seemingly no distinct 'cervico-auriculo-occipitalis' in the orangutans dissected by us. For instance, in our HU PP1 specimen the occipitalis has a superficial bundle (Fig. 4) that originates near the midline, mainly from an aponeurosis, its inferior fibers being more horizontal and running to the posterior margin of the ear (being always deeper to the auricularis superior, contra Sullivan & Osgood 1925), while its superior fibers are more vertical, running to the galea aponeurotica and the frontalis. The deeper bundle (Fig. 4) originates 1 or 2 cm laterally to the midline, at about the level of the nuchal line, its fibers running transversally (perpendicularly to those of the superficial bundle) to reach the midline superiorly, meeting with its counterpart. The orientation of the inferior portion of this deep bundle is somewhat similar to that of the sternocleidomastoideus, as described by Sullivan & Osgood (1925). As also described by Sullivan & Osgood (1925) the deep bundle lies just superiorly to the trapezius, and does not overlap with this latter muscle, and the superior/mesial portion of this deep bundle then fuses with the deep fascia of the scalp, its fleshy portion having only about 4 cm of total length. In this HU PP1 specimen, the occipitalis may possibly include the **auricularis posterior**, which is usually not present as a distinct muscle in orangutans; however, we have no strong evidence to support that the latter muscle is included in the occipitalis. Sonntag (1924a), Lightoller (1928a), Seiler (1976) and Winkler (1989) stated that some orangutans have an 'auricularis posterior', but the descriptions and illustrations of these authors seem to indicate that this structure is actually deeply blended to the occipitalis (e.g., Fig. 1 of Sonntag 1924a), so it is not clear if these orangutans do have a distinct muscle auricularis posterior *sensu* the present work. According to Ruge (1887a,b), Sullivan & Osgood (1925), Huber (1930a,b, 1931), Lightoller (1928a) and Miller (1952), as well as to our dissections, the auricularis posterior is in fact not present as a distinct muscle in most *Pongo* specimens. As explained above, the 'cervico-auriculo-occipitalis' is seemingly missing in this HU PP1 specimen. The muscle **mandibulo-auricularis** is usually also not present as a distinct structure in orangutans.

- Synonymy: Part or totality of auriculo-occipitalis (Sullivan & Osgood 1925, Lightoller 1928a) and of auriculo-occipitalis proprius (Lightoller 1928a), which includes the occipitalis + posterior auricularis *sensu* the present work; part or totality of occipito-auricularis (Edgeworth 1935).

Intrinsic facial muscles of ear
- Usual attachments: See notes below.
- Usual innervation: Branches of CN7.
- Notes: The intrinsic facial muscles of the ear of orangutans have been rarely described in detail in the literature, and they were difficult to analyze in our specimens. However, a few authors such as Seiler (1976) have examined these muscles in some detail. According to Seiler (1976) the **helicis**, **antitragicus**, **tragicus**, **obliquus auriculae** and **transversus auriculae** are usually present in orangutans, as is normally the case in modern humans. Contrary to Ruge 1887a,b, authors such as Huber (1930b, 1931) and Richmond (1993) also described vestigial muscles 'auriculares proprii' ('musculus auriculae proprius posterior' *sensu* Ruge 1887a,b, which includes the obliquus auriculae and/or transversus auriculae sensu the present work) in *Pongo*. Seiler (1976) suggested that the **incisurae terminalis** ('**incisurae Santorini**'), **depressor helicis** and the '**intercartilagineus**' are usually absent in orangutans, and that the **pyramidalis auriculae** ('**trago-helicinus**' *sensu* Edgeworth 1935 and Seiler 1974a) is usually reduced to connective tissue in *Pongo*, suggesting that in this taxon there is no distinct, fleshy muscle pyramidalis auriculae.

Zygomaticus major (Fig. 1)
- Usual attachments: From the temporalis fascia and the zygomatic arch/bone (not from the ear, although it often originates nearer to the ear than is usually the case in modern humans) to the corner of the mouth, passing superficial to the levator anguli oris facialis.
- Usual innervation: Branches of CN7.
- Notes: The 'orbito-labialis' *sensu* Sullivan & Osgood (1925) corresponds, according to these authors, to the zygomaticus minor *sensu* the present study, but in fact it seems to correspond to the zygomaticus major + zygomaticus minor; therefore, when they state that the 'zygomaticus major' is missing in *Pongo*, what this means is that this muscle is not well separated from the zygomaticus minor. It could seem that the zygomaticus' and the 'caput zygomaticus of the quadratus labii superioris' *sensu* Lightoller (1928a) and Loth (1931) would correspond, respectively, to the zygomaticus major and zygomaticus minor *sensu* the present study. However, it is in fact the structure that Lightoller (1928a) designates as 'malaris' (or 'pars peripheralis of the orbicularis oris') in *Pongo* that seems to correspond to the zygomaticus major and zygomaticus minor *sensu* the present study. Saban (1968, p. 530) confirms that the 'orbito-labialis' of *Pongo* (*sensu* Sullivan & Osgood 1925) corresponds to the 'auriculo-labialis' *sensu* Lightoller (1928a), which corresponds to the zygomaticus major and/or zygomaticus minor of *Homo*. The **depressor tarsi** ('**preorbicularis**') is usually not present as a distinct muscle in orangutans. Seiler (1971d, 1976)

suggested that a **'risorius'** might be occasionally present in *Hylobates* and *Pongo*, but as explained by Diogo & Wood (2011, 2012), at least some of the structures shown by this author in specimens of these genera clearly do not seem to be homologous to each other, nor to the risorius of *Homo*, *Pan* and *Gorilla*. In fact, at least to our knowledge all the other authors that have studied in detail the facial muscles of *Hylobates* and *Pongo* did not found a distinct muscle risorius in any specimens of these two genera; moreover, we did not also found this muscle in any of the hylobatid and *Pongo* specimens dissected by us. Therefore, even if some of the structures described in these two taxa by Seiler (1971d, 1976) are actually homologous to the risorius of *Homo*, *Gorilla* and *Pan*, the fact is that, contrarily to what happens in these latter genera, the presence of a risorius would, anyway nevertheless still represent an extremely rare condition within Asian apes.

- Synonymy: Seems to correspond to part of the zygomaticus major *sensu* Owen (1830-1831), of the zygomatic mass *sensu* Sonntag (1924a), of the orbito-labialis *sensu* Sullivan & Osgood (1925) and of the malaris, or pars peripheralis of the orbicularis oculi, *sensu* Lightoller (1928a); zygomaticus (Ruge 1887a,b); mainly corresponds to zygomaticus inferior *sensu* Seiler (1976).

Zygomaticus minor (Fig. 1)
- Usual attachments: Mainly from the zygomatic bone and the orbicularis oculi (relatively far from the ear and near to the eye, as is usually the case in modern humans: see, e.g., plate 26 of Netter 2006) to the corner of the mouth and to the upper lip, being mainly superficial to the levator anguli oris facialis.
- Usual innervation: Branches of CN7.
- Notes: See notes about zygomaticus major, above.
- Synonymy: Seems to correspond to part of the zygomaticus major *sensu* Owen (1830-1831), of the zygomatic mass *sensu* Sonntag (1924b), of the orbito-labialis *sensu* Sullivan & Osgood (1925) and of the malaris, or pars peripheralis of the orbicularis oculi, *sensu* Lightoller (1928a); mainly corresponds to zygomaticus superior *sensu* Seiler (1976).

Frontalis (Fig. 1)
- Usual attachments: From the galea aponeurotica to the skin of the eyebrow and nose.
- Usual innervation: Branches of CN7.
- Function: See platysma myoides above.
- Synonymy: Orbito-auricularis (Edgeworth 1935).

Auriculo-orbitalis (Figs. 1–2)
- Usual attachments: From the anterior portion of ear to the region of the frontalis.
- Usual innervation: Temporal branches of CN7.

• Notes: In Terminologia Anatomica (1998) the **temporoparietalis** is considered to be a muscle that is usually present in modern humans (originating mainly from the lateral part of the galea aponeurotica and passing inferiorly to insert onto the cartilage of the auricle, in an aponeurosis shared with the other auricular muscles). However, according to authors such as Loth (1931) the temporoparietalis is usually absent as a distinct muscle in modern humans. According to Diogo et al. (2008, 2009b), the temporoparietalis and **auricularis anterior** derive from the auriculo-orbitalis so when the temporoparietalis is not present as a distinct muscle these authors use the name auriculo-orbitalis to designate the structure that is often designated in the literature as 'auricularis anterior': that is, one can only use this latter name when the temporoparietalis is present. In various primates reported by Seiler (1976) including hominoids such as *H. moloch* and *Pongo pygmaeus*, he reported both a 'pars orbito-temporalis of the frontalis' and an 'auricularis anterior' attaching posteriorly onto the ear. That is, the structure that he designated as 'auricularis anterior' is differentiated from the auriculo-orbitalis, as is the case in various other primates, but according to him, and contrary to the condition in *Pan troglodytes* and *Gorilla gorilla* (e.g., Fig. 143 of Seiler 1976), in hominoids such as *H. moloch* and *P. pygmaeus* there is 'still' a connection between the main body of the auriculo-orbitalis and the ear. One would think that the temporoparietalis of taxa such as modern humans would probably correspond to those remaining fibers of the auriculo-orbitalis that did not differentiate into the 'auricularis anterior', and this was what Jouffroy & Saban (1971) suggested in their study on the facial muscles of mammals. However, in at least some, if not all, taxa this is clearly not the case. For example, in Fig. 143 of Seiler (1976) the structure that he designates as 'pot', which corresponds to the remaining fibers of the auriculo-orbitalis that do not form an 'auricularis anterior', does not correspond to the structure that is usually designated as temporoparietalis in modern human anatomical atlases, which as its name indicates, usually runs mainly superoinferiorly from the parietal bone to the temporal region. Therefore, in order to be as consistent as possible with our previous studies and because the 'pars orbito-temporalis of the frontalis' and the 'auricularis anterior' *sensu* Seiler (1976) are very likely derived from the same anlage and are often related to each other, being often even continuous, we consider these two structures as parts/bundles of the auriculo-orbitalis *sensu* the present study (see, e.g., Fig. 74 of Seiler 1976). That is, the 'auricularis anterior' *sensu* Seiler (1976) is considered to be part of the auriculo-labialis *sensu* the present study, except in those few primates that have a distinct temporoparietalis (in those few primates the 'auricularis anterior' *sensu* Seiler 1976 is named by us as auricularis anterior, to clearly indicate that those primates have a distinct temporoparietalis). It is however possible, and in our opinion likely, that as suggested by Jouffroy & Saban (1971) the temporoparietalis of modern humans corresponds to the 'pars orbito-temporalis of the frontalis' *sensu* Seiler (1976) and that most of the confusion related to this issue is due to erroneous descriptions of the modern human temporoparietalis in anatomical atlases (which suggest that this is mainly a vertical muscle running superoinferiorly from the parietal bone to the temporal region, a description that does not match with the usual configuration of Seiler's 'pars orbito-temporalis of the frontalis' in other primates, which mainly runs horizontally to the region of the orbit to the region of the ear). If this is the case, then most primates would have a distinct temporoparietalis and a distinct auricularis anterior because both a 'pars orbito-temporalis of the frontalis' and an 'auricularis anterior' were reported in most primates by Seiler (1976). If this is so, then the 'pars orbito-temporalis of the frontalis' and the 'auricularis anterior' of those primates should be designated as temporoparietalis and as auricularis anterior, respectively; we plan to return to this subject in the future. In our HU PP1 specimen (Fig. 1) the temporoparietalis and the auricularis anterior are not present as separate muscles, so the auriculo-orbitalis of *Pongo* could well correspond to the temporoparietalis plus auricularis anterior of humans. The anterior fibers of the auriculo-orbitalis attach onto the region of the superciliary ridge and are somewhat blended with those of the frontalis, but its posterior fibers pass mesially to, and are separated from, those of the frontalis, attaching onto the anterior margin of the ear. The **zygomatico-auricularis** is not present as a distinct muscle is orangutans.

- Synonymy: Probably corresponds to part of the frontalis *sensu* Owen (1830-1831); orbitotemporalis or orbitoauricularis (Lightoller 1928a, Jouffroy & Saban 1971, Winkler 1989).

Auricularis superior (Figs. 1–2, 4)

- Usual attachments: From the superior margin of the ear to the galea aponeurotica and at least in some cases also the temporal fascia (Gibbs 1999).
- Usual innervation: Branches of CN7.
- Notes: Huber (1930a, 1931) stated that the muscle that is often named 'auricularis superior' in *Pongo* is not homologous to the auricularis superior of other primates, because it is a post-auricular muscle (i.e., derived from the 'platysma') and not a post-auricular muscle (i.e., derived from the sphincter colli profundus), as is the auricularis superior of other primates. According to him, the 'auricularis superior' of *Pongo* extends more superiorly than the auricularis superior of other primates, and some fibers of the 'true' auricularis superior may be recognized in *Pongo*, but they are deeply blended with those of the frontalis. However, our dissections as well as the dissections of various other authors (e.g., Sonntag 1924a; Lightoller 1928a, Winkler 1989) clearly indicate that orangutans have a well-developed auricularis superior muscle that is similar, and homologous, to that of other primates (Figs. 1–2, 4).

Orbicularis oculi (Fig. 1)
- Usual attachments: From a continuous bony attachment around the orbit and also from the medial palpebral ligament, to skin near the eye; the muscle is usually divided into a pars palpebralis and a pars orbitalis, as in modern humans, and in at least some cases it sends a tendon to the zygomatic bone.
- Usual innervation: Branches of CN7.
- Notes: The orbicularis oculi of orangutans is often differentiated into a pars orbitalis and a pars palpebralis, which is in turn subdivided into a pars profunda ('pars lacrimalis') and a fasciculus ciliaris (e.g., Sullivan & Osgood 1925, Lightoller 1928a, Winkler 1989; *contra* Sonntag 1924a, who stated that none of these structures is separated in *Pongo*). In our HU PP1 specimen the muscle is exactly as described by Sullivan & Osgood (1925), the structure designated as 'pars orbicularis malaris' by these authors seemingly being a portion of the orbicularis oculi that superiorly is somewhat similar to the zygomaticus minor, but that inferiorly is very distant from the normal origin of the zygomaticus minor. This 'pars orbicularis malaris' is highly blended with a muscle that lies posteriorly and laterally to it, which corresponds to the zygomaticus major. We could not check if the orbicularis oculi has an origin from the posterior lacrimal crest, but we did find the tendon of this muscle to the zygomatic bone. Seiler (1971d, 1976) states that, within the Catarrhini, an independent muscle '**infraorbitalis**' is present in *Macaca*, *Pan* and *Homo* and occasionally in *Pongo*; however, this structure seems to correspond to part of the orbicularis oculi and/or of the levator labii superioris alaeque nasi *sensu* the present study. The **zygomatico-orbicularis** is missing in orangutans.
- Synonymy: Orbicularis palpebrarum (Owen 1830-1831).

Depressor supercilii (Fig. 1)
- Usual attachments: From the ligamentum palpebrale mediale and in at least some cases also from the nasal process of the nasal bone, to the eyebrow and often also to the supraorbital fascia (Gibbs 1999).
- Usual innervation: Branches of CN7.
- Notes: In our HU PP1 specimen the depressor supercilii has superficial (broad) and deep (thinner) portions, which are separated inferiorly by fibers of the orbicularis oculi, as shown in the dioptogram A1 of Lightoller (1928a).
- Synonymy: Depressor capitis (Lightoller 1928a); retractor anguli oculi medialis (Jouffroy & Saban 1971).

Corrugator supercilii (Fig. 1)
- Usual attachments: From the orbital margin and glabella to the eyebrow region.
- Usual innervation: Branches of CN7.
- Notes: Gibbs (1999) suggested that the corrugator supercilii is absent in 1/2 *Pongo* and some authors did not describe the muscle in this genus (e.g., Owen 1830-1831,

Sonntag 1924a), but in the specimens dissected by us and by authors of studies focused specifically on facial musculature (e.g., Lightoller 1928a, Huber 1930b, 1931, Seiler 1976), the muscle was consistently present and well-developed.

Levator labii superioris (Figs. 1–2)

- Usual attachments: From the infraorbital region mainly to the upper lip.
- Usual innervation: Branches of CN7.
- Synonymy: Part or totality of the maxillo-labialis (Sullivan & Osgood 1925) and of the infraorbital head of the quadratus labii superioris (Lightoller 1928a, Jouffroy & Saban 1971).

Levator labii superioris alaeque nasi (Fig. 1)

- Usual attachments: From the region of the ligamentum infraorbitale mediale, infraorbital margin and/or maxilla to the upper lip and sometimes to the ala of the nose (e.g., Lightoller 1928a, Gibbs 1999).
- Usual innervation: Branches of CN7.
- Synonymy: Levator alae nasi (Sonntag 1924a); part or totality of the maxillo-labialis (Sullivan & Osgood 1925); angular head of the quadratus labii superioris (Lightoller 1928a, Jouffroy & Saban 1971).

Procerus (Fig. 1)

- Usual attachments: From the frontalis to the dorsomedial region of the nose.
- Usual innervation: Branches of CN7.
- Notes: According to Seiler (1971c, 1976) the '**depressor glabellae**' is usually present as a distinct muscle in orangutans, whereas the procerus is inconstant. However, the 'depressor glabellae' is often considered in the literature to be part of the procerus (see, e.g., Terminologia Anatomica 1998). Some authors did not describe the procerus in *Pongo* (e.g., Sonntag 1924a), but in the specimens dissected by us and by authors of studies focused specifically on facial musculature (e.g., Huber 1930b, 1931, Seiler 1976) the muscle was consistently present, although it is often blended with surrounding muscles such as the levator labii superioris alaeque nasi and/or the frontalis.
- Synonymy: Procerus nasi or depressor glabellae (Huber 1930b, 1931); procerus plus part or totality of depressor glabellae (Seiler 1971c, 1976).

Buccinatorius (Figs. 2, 5, 22)

- Usual attachments: From the pterygomandibular raphe, the infero-lateral surface of the maxilla, and the supero-lateral border of the mandible, mainly to the angle of the mouth and the upper and lower lips.
- Usual innervation: Branches of CN7.

Nasalis (Figs. 1–2)

- Usual attachments: From the maxilla, deep to the orbicularis oris, to the lateral margin of the nose and the lateral portion of the inferior margin of the ala of the nose.
- Usual innervation: Branches of CN7.

- Notes: Seiler (1970, 1971c, 1976) describes a 'nasalis' and a **'subnasalis'** in various catarrhines. The 'subnasalis' and 'nasalis' *sensu* Seiler could correspond to the pars alaris and pars transversa of the nasalis of modern human anatomy, respectively (compare, e.g., Fig. 141 of Seiler 1976 to plate 26 of Netter 2006). However, according to Seiler the 'subnasalis' while often present in orangutans and chimpanzees is usually missing in *Homo* and *Gorilla*. This indicates that the 'subnasalis' does not correspond to the pars alaris of the nasalis *sensu* the present study, because this latter structure *is* usually found in modern humans. Be that as it may, as other authors do not refer to a 'subnasalis' muscle and as we also did not find a distinct separate 'subnasalis' in the orangutans dissected by us, this structure probably corresponds to part of the nasalis and/or of the orbicularis oris *sensu* the present study. Seiler (1970, 1971c, 1976) also describes a **'depressor septi nasi'** and a **'musculus nasalis impar'** in a few catarrhines, but not in orangutans. One hypothesis is that the 'nasalis impar' corresponds to part or to the totality of the depressor septi nasi that is illustrated in a few atlases of modern human anatomy as a vertical muscle that lies on the midline and that attaches mainly onto the inferomesial margin of the nose (compare Fig. 145 of Seiler 1976 with plate 26 of Netter 2006). However, most atlases of modern human anatomy show two depressor septi nasi muscles, one in each side of the body, running obliquely (superomedially) from the upper lip to a more medial part of the inferior region of the nose. Thus, the 'nasalis impar' *sensu* Seiler (1976) probably corresponds to an additional midline muscle that is only inconstantly present in catarrhines, while the 'depressor septi nasi' *sensu* Seiler (1976) is effectively similar to the depressor septi nasi shown in most atlases of modern human anatomy. Table 3 of Seiler (1979b) refers to a **'labialis superior profundus'** and this suggests that according to him the 'depressor septi nasi' and the 'labialis superior profundus' are not the same structure, whereas Lightoller (1928a, 1934) suggests that these are one and the same structure (Seiler 1976 clearly shows both these structures in various primates, e.g., in his Fig. 145 of *Gorilla*, the 'labialis superior profundus' probably corresponding to part of the orbicularis oris *sensu* the present study). Jouffroy & Saban (1971) designate the nasalis as the 'naso-labialis profundus pars anterior' and the depressor septi nasi as the 'naso-labialis profundus pars mediana'; this suggests that these authors consider these two muscles derive from the same structure. In their Fig. 471, Jouffroy & Saban (1971) designate the 'labii profundus superioris' as depressor septi nasi, thus suggesting that the 'labii profundus superior' could correspond to the depressor septi nasi of modern humans as suggested by Lightoller (1928a, 1934). However, our comparisons and dissections suggest that the homologies proposed by Seiler (1976) are somewhat doubtful and that: 1) the 'nasalis' *sensu* Seiler likely corresponds to part or the totality of the nasalis *sensu* the present study; 2) the pars transversa of the nasalis of modern humans might correspond to the 'subnasalis' *sensu* Seiler (see, e.g., Fig. 141 of Seiler 1976; however, Seiler stated that the 'subnasalis' is missing in modern humans), to the depressor septi nasi *sensu* Seiler 1976 (see, e.g., Fig. 145 of Seiler 1976), and/or to part of the nasalis *sensu* the present study (see, e.g., Fig. 145 of Seiler 1976); 3) the depressor septi nasi of modern humans corresponds to the 'depressor septi' nasi *sensu* Seiler; therefore, the 'depressor septi nasi' *sensu* Seiler does not correspond to part of the 'labialis superior profundus' *sensu* Seiler, as suggested by Lightoller (1928a, 1934; see, e.g., Fig. 141 of Seiler 1976), nor to the 'nasalis impar' *sensu* Seiler (in fact, Seiler stated that the 'nasalis impar' is missing in modern humans, which do usually have a depressor septi nasi *sensu* the present study).

Depressor septi nasi (Figs. 1–2)

- Usual attachments: From the maxilla, deep to the orbicularis oris, to the inferior region of the nose.
- Usual innervation: Branches of CN7.
- Function: Lightoller (1928a) stated that in the three orangutans dissected by him the 'pars perpendicularis' of the 'nasalis' (or 'labii profundus superior', which might correspond to the depressor septi nasi *sensu* the present study) would seemingly drag caudally (orally) the medial portion of the orbicularis oris together with the loose tissue inferior to the nares, and depress the lip margin; when acting with the other labial muscles (e.g., the orbicularis oris) the lip would be everted owing to the dynamic force of a couple.
- Notes: See nasalis, above.
- Synonymy: Pars perpendicularis of nasalis or labii profundus superior (Lightoller 1928a).

Levator anguli oris facialis (Fig. 2)

- Usual attachments: From the canine fossa of the maxilla to the angle of mouth.
- Usual innervation: Branches of CN7.
- Function: Sullivan & Osgood (1925) stated that in orangutans the levator anguli oris facialis and the depressor anguli oris pull the corner of the mouth forward, and the levator anguli oris facialis also raises the corner of the mouth and the lower lip.
- Notes: As proposed by Diogo et al. (2008) and Diogo & Abdala (2010), we use the name 'levator anguli oris facialis' here (and not the name 'levator anguli oris', as is usual in atlases of modern human anatomy) to distinguish this muscle from the **levator anguli oris mandibularis** (which is usually also designated as 'levator anguli oris' in the literature) found in certain reptiles, which is part of the mandibular (innervated by CN5) and not of the hyoid (innervated by CN7) musculature. The **anomalus nasi** and **anomalus menti**, found as anomalies in a few modern humans, are usually missing in orangutans. Lightoller (1928a) reported a *Pongo* specimen with some transverse fibers close to the canine fossa, which constituted an '**anomalus maxillae**' and were immediately lateral to the

origin of the levator anguli oris facialis and extended from the canine fossa to the malar eminence. According to him, the position of this 'anomalus maxillae' was an unusual one and the fibers probably were derived from the anlage that gives rise to the levator anguli oris facialis.

- Synonymy: Levator anguli oris (Owen 1830-1831; Sonntag 1924a); caninus (Lightoller 1928a); maxillo-labialis, depressor labi communis, or pyramidalis menti (Jouffroy & Saban 1971); part of caninus (Seiler 1976), which also includes the depressor anguli oris *sensu* the present work.

Orbicularis oris (Figs. 1–2)

- Usual attachments: From skin, fascia and adjacent regions of lips to skin and fascia of lips.
- Usual innervation: Branches of CN7.
- Notes: Seiler (1970, 1971c) describes a '**cuspidator oris**' in orangutans. This structure, which was designated as '**labialis superior profundus**' by Seiler 1976, probably corresponds to the '**incisivus labii superior**' *sensu* Lightoller (1928, 1934, 1939) and, thus to part of the orbicularis oris *sensu* the present study. Seiler (1976) also describes a '**labialis inferior profundus**' in orangutans, which thus probably corresponds to the '**incisivus labii inferioris**' *sensu* Lightoller (1928, 1934, 1939) and to part of the orbicularis oris *sensu* the present study.
- Synonymy: Orbicularis oris plus incisivus superior and incisivus inferior (Lightoller 1928a).

Depressor labii inferioris (Figs. 1–2)

- Usual attachments: From the platysma myoides and the mandible to the lower lip.
- Usual innervation: Branches of CN7.
- Synonymy: Quadratus labii inferioris (Sonntag 1924a, Sullivan & Osgood 1925, Lightoller 1928a); buccinatorius pars mandibularis, quadratus menti, or mento-labialis (Jouffroy & Saban 1971).

Depressor anguli oris (Figs. 1–2)

- Usual attachments: From the angle of mouth to the fascia of the platysma myoides.
- Usual innervation: Branches of CN7.
- Notes: There is usually no **transversus menti** in orangutans.
- Synonymy: Triangularis (Sonntag 1924a, Lightoller 1928a, Edgeworth 1935); part of caninus *sensu* Seiler 1976, which also includes the levator anguli oris facialis *sensu* the present work.

Mentalis

- Usual attachments: From the mandible to the skin below the lower lip, being mainly deep to the orbicularis oris and to the depressor labii inferioris, sometimes meeting with its counterpart in the midline (e.g., Lightoller 1928a, Gibbs 1999, our dissections).

- Usual innervation: Branches of CN7.
- Function: Lightoller (1928a) stated that in orangutans the mentalis lifts orally the medial portion of the orbicularis oris together with the loose tissues of the chin, and hence forces cranially (orally) the lower lip.
- Synonymy: Labii profundus inferior (Lightoller 1928a); levator labii inferioris, levator menti, incisivus labii inferioris, or incisivus mandibularis (Jouffroy & Saban 1971).

3.3 Branchial musculature

Stylopharyngeus
- Usual attachments: Mainly from the styloid process to the pharyngeal wall (e.g., Dean 1984; Gibbs 1999), passing between the middle and superior pharyngeal constrictors.
- Usual innervation: Data are not available.
- Notes: The **ceratohyoideus** and **petropharyngeus** are not present as distinct muscles in orangutans.

Trapezius (VU PP1 LSB = 307.9 g; Figs. 13, 25–26, 29, 35, 37)
- Usual attachments: Mainly from the vertebral column and cranium to the spine and acromion of the scapula and to the lateral 1/3, or more than the lateral 1/3, of the clavicle (NB the ligamentum nuchae is usually not present as a distinct, well-defined structure in orangutans).
- Usual innervation: Spinal accessory nerve (Primrose 1899, 1900); accessory nerve plus C3 (Schück 1913b, Kallner 1956).
- Function: Tuttle & Basmajian (1976) stated that the pars ascendens of the trapezius exhibited EMG activity during arm-raising in *Pongo*.
- Notes: As in modern humans, in orangutans the trapezius is usually differentiated into a pars descendens, a pars transversa and a pars ascendens. In *Pongo* the trapezius usually goes to the lateral 1/3, or to more than the lateral 1/3, of the clavicle. This is the case in all *Pongo* specimens dissected by us and reported in the literature we reviewed, except a specimen reported by Hepburn (1892) in which the attachment was to the acromial end of the clavicle and a specimen described by Schück (1913a,b) in which the trapezius attached exclusively onto the scapular spine and acromion (i.e., not onto the clavicle).

Sternocleidomastoideus (VU PP1 LSB = 123.6 g; Figs. 4–6, 8, 25–26)
- Usual attachments: The caput sternomastoideum runs mainly from the sternum to the mastoid region and the lateral portion of the superior nuchal line; the caput cleidomastoideum runs mainly from the medial portion of the clavicle (medial 1/4: e.g., Primrose 1899, 1900; medial 1/3: Sonntag 1924a) to the mastoid region.
- Usual innervation: Accessory nerve (Schück 1913b); in the male orangutan dissected by Kallner (1956) innervation was by the accessory nerve, as it was on

one side of the female orangutan she examined; however, on the other side of the female orangutan innervation was by the accessory nerve and by C2.

- Notes: In a few orangutans there is a structure that is sometimes designated as muscle '**cleido-occipitalis**' (e.g., Sonntag 1924a) but this structure is usually deeply blended with the sternocleidomastoideus, and does not form a distinct, separate, muscle, i.e., it is instead a bundle of the sternocleidomastoideus, which is designated in the present work as **caput cleido-occipitale**. According to Wood (1870) and to the recent work of Mustafa (2006) only about 36% and 33% of modern humans, respectively, have a caput cleido-occipitale of the sternocleidomastoideus. In one *Pongo* specimen dissected by Owen (1868) the caput cleidomastoideum inserted onto the 'diapophysis of the axis vertebra'.
- Synonymy: Sternomastoid (Primrose 1899, 1900); sternomastoid plus cleidooccipital (Sonntag 1924a).

Constrictor pharyngis medius
- Usual attachments: From the dorsal raphe of the pharyngeal wall to hyoid bone (e.g., Sonntag 1924a).
- Usual innervation: Data are not available.

Constrictor pharyngis inferior
- Usual attachments: From the dorsal raphe of the pharyngeal wall to the thyroid cartilage (pars thyropharyngea), the cricoid (pars cricopharyngea) cartilage.
- Usual innervation: Data are not available.

Cricothyroideus
- Usual attachments: From the cricoid cartilage to the thyroid cartilage.
- Usual innervation: Data not available.
- Notes: Apart from a pars recta and a pars obliqua, a pars interna is present in *Pongo* according to Kohlbrügge (1896), Duckworth (1912), and Saban (1968), but not according to Brandes (1932) and Starck & Schneider (1960). The *thyroideus transversus* is usually not present in orangutans.
- Synonymy: Cricothyreoideus anticus (Kohlbrügge 1896); thryro-cricoideus or thyroideus transversus (Saban 1968).

Constrictor pharyngis superior
- Usual attachments: It originates from the dorsal raphe of the pharyngeal wall and from the basioccipital (e.g., Dean 1985). It is however not clear if the pars mylopharyngea, pars pterygopharyngea and/or pars buccopharyngea of the constrictor pharyngis superior are usually present, or not, in *Pongo*; the pars glossopharyngea is present in this taxon according to Kleinschmidt (1938).
- Usual innervation: Data are not available.
- Notes: We could not discern, in the specimens dissected by us, if the **pterygopharyngeus**, **salpingopharyngeus** and/or **patatopharyngeus** are usually present or not as a distinct muscle in orangutans.

Musculus uvulae

- Usual attachments: Authors such as Hill (1939) stated that in *Pongo* the uvula is usually absent, and therefore that its 'azygos muscle' is also missing. However, the musculus uvulae has been described in orangutans by various other authors (e.g., Chapman 1880; Sonntag 1924a,b), including Jordan (1971a,b,c), who stated the musculus uvulae is present in this taxon, but may not enter the uvula.
- Usual innervation: Data are not available.
- Synonymy: Azygos uvulae (Chapman 1880; Sonntag 1924a).

Levator veli palatini

- Usual attachments: Dean (1985) reported an origin from the petrous apex and an insertion onto the superior surface of the soft palate.
- Usual innervation: Data are not available.
- Synonymy: Levator palatini (Sonntag 1924a); pterygo-palatinus (Saban 1968).

Thyroarytenoideus

- Usual attachments: See notes below.
- Usual innervation: Inferior laryngeal nerve (from recurrent laryngeal nerve) (Hill 1939).
- Notes: With respect to hominoids and other primates there has been controversy regarding the homologies of the thyroarytenoid bundles and the presence/absence of a distinct **musculus vocalis**. Kohlbrügge (1896) dissected gorillas, chimpanzees and orangutans as well as taxa such as *Cebus, Semnopithecus, Hylobates* and *Macaca*, and stated that he could not find a distinct attachment of the thyroarytenoideus onto a true vocal cord (such as that found in modern humans) in any of these taxa, except perhaps in *Pongo*; within all the taxa mentioned above, he found an attachment onto the cricoid cartilage in *Hylobates* and *Colobus*. Duckworth (1912) examined specimens from all of the five extant hominoid genera, as well as from cadavers belonging to *Macaca, Cebus, Semnopithecus* and *Tarsius*, and suggested that a well-developed, distinct musculus vocalis associated with the plica vocalis is also only consistently present in modern humans. But Duckworth did suggest that great apes, and particularly chimpanzees, show a configuration that is somewhat similar to that found in modern humans, in that they have a poorly developed/differentiated musculus vocalis (see, e.g., his Figs. 24 and 17). Edgeworth (1935) defended the notion that a musculus vocalis is found in some non-human primates and suggested that when this structure is present the thyroarytenoideus becomes a **'thryroarytenoideus lateralis'** because its inferior/mesial part gives rise to the vocalis muscle. Starck & Schneider (1960) described a 'pars lateralis' as well as a 'pars medialis' of the musculus vocalis that usually goes to the vocal fold/cord in *Pan, Pongo* and *Gorilla*, but not in *Hylobates*; the latter corresponds to the musculus vocalis of modern humans. They did not find a pars aryepiglottica

or a pars thyroepiglottica of the thyroarytenoideus in *Hylobates* and *Pongo*, but stated that other authors did report at least one of these structures in *Gorilla* and *Pan*. Saban (1968) clarified the nomenclature of the thyroarytenoideus and suggested that this muscle may be divided into the following structures: 1) a pars superior (often designated as '**thyroarytenoideus superior**', 'thyroarytenoideus lateralis' or '**ventricularis**'); 2) a pars inferior (often designated as '**thyroarytenoideus inferior**', '**thyroarytenoideus medialis**', or musculus vocalis); 3) a '**ceratoarytenoideus lateralis**; 4) a 'pars intermedia' (but this name was only used by a few authors such as Starck & Schneider 1960 who stated that some primates might have a pars superior, a pars inferior, and a pars intermedia); 5) a pars thyroepiglottica; 6) a pars aryepiglottica; 7) a pars arymembranosa; and 8) a pars thyromembranosa. According to Saban (1968) the 'ceratoarytenoideus lateralis' is usually fused with (but not differentiated from, as suggested in some anatomical atlases) the cricoarytenoideus posterior, being only a distinct muscle in a few taxa and, within primates, in *Pan* (and still in this case this seems to constitute a variant/anomaly) where it is a small muscle running from the dorsal face of the inferior thyrohyoid horn to the arytenoid cartilage. Also according to Saban the pars superior and pars inferior are more superior and inferior, respectively, in apes and modern humans (in modern humans the more inferior or medial part, i.e., the pars inferior is well developed and often designated as musculus vocalis), whereas in primates such as *Macaca* and *Papio* they are more lateral/medial; in *Pongo* the pars inferior is well-developed and lies anterior to the vocal cord, but it is not associated with it. Aiello & Dean (1990) stated that in non-human hominoids the pars aryepiglottica is often reduced in size or absent. Our observations and review of the literature indicate that the pars superior and pars inferior of the thyroarytenoideus are usually present in orangutans. The ceratoarytenoideus lateralis, pars intermedia, **pars thyroepiglottica**, **pars aryepiglottica**, **pars thyromembranosa** and **pars arymembranosa** are usually not present in orangutans according to Starck & Schneider (1960).
- Synonymy: Thyroarytenoideus plus musculus vocalis (Duckworth 1912, Jordan 1971a,b,c).

Musculus vocalis (see thyroarytenoideus above)
- Usual attachments: See thyroarytenoideus above.
- Usual innervation: See thyroarytenoideus above.
- Notes: See thyroarytenoideus above.

Cricoarytenoideus lateralis
- Usual attachments: From the anterior portion of the cricoid cartilage to the muscular process of the arytenoid cartilage.
- Usual innervation: Data are not available.

Arytenoideus transversus
- Usual attachments: Unpaired, transversly-oriented muscle connecting the two contralateral arytenoid cartilages.
- Usual innervation: Data are not available.

Arytenoideus obliquus
- Usual attachments: From the arytenoid cartilage to the contralteral arytenoid cartilage; its fibers are more oblique than those of the arytenoideus transversus.
- Usual innervation: Data are not available.
- Notes: The arytenoideus obliquus is present as a distinct muscle in at least some orangutans because it was reported for instance by Duvernoy (1855-1856), Fürbringer (1875) and Kleinschmidt (1938), although the muscle was not reported in *Pongo* by various other authors, e.g., Sonntag (1924a) and Starck & Schneider (1960).

Cricoarytenoideus posterior
- Usual attachments: From dorsal portion of the cricoid cartilage to the arytenoid cartilage; in at least some cases the muscle meets its counterpart at the dorsal middline (e.g., Sonntag 1924a).
- Usual innervation: Data are not available.
- Notes: There is seemingly no attachment of the muscle onto the inferior horn of the thyroid cartilage in orangutans; thus there is seemingly no **ceratocricoideus**.
- Synonymy: Cricoarytenoideus posticus (Kohlbrügge 1896, Hill 1939).

3.4 Hypobranchial musculature

Geniohyoideus
- Usual attachments: From the mandible to the hyoid bone, lying close to, but usually not merging with, its counterpart at the midline.
- Usual innervation: Hypoglossal nerve (Sonntag 1924a); ventral ramus of the first cervical nerve through the hypoglossal nerve (Wall et al. 1994).
- Function: Stimulation of the geniohyoideus in *Pongo* elicited slight mandibular depression and tongue protrusion (Waller et al. 1994).

Genioglossus
- Usual attachments: From the mandible to the tongue and also to the hyoid bone (e.g., Edgeworth 1935 referred to the fibers that attach onto the hyoid bone as part of a '**genio-hyoglossus**'). The genioglossus muscles are usually well-separated by a midline fibrous septum and/or fatty tissue (e.g., Saban 1968).
- Usual innervation: Hypoglossal nerve (Sonntag 1924a).

- Notes: The muscles **genio-epiglotticus**, **glosso-epiglotticus**, **hyo-epiglotticus**, and **genio-hyo-epiglotticus**, described by authors such as Edgeworth (1935) and Saban (1968) in some primate and non-primate mammals, do not seem to be usually present as distinct structures in orangutans (according to these authors these muscles are usually not present in catarrhine taxa).

Intrinsic muscles of tongue
- Usual attachments: To our knowledge, there are no detailed published descriptions of these muscles in orangutans and we could not examine them in detail in our dissections. However, the **longitudinalis superior**, **longitudinalis inferior**, **transversus linguae** and **verticalis linguae** are consistently found in modern humans and at least some other primate and non-primate mammals, so these four muscles are very likely also present in orangutans, and at least the longitudinalis inferior was reported in *Pongo* by Hill 1939. Detailed studies of the tongue and its muscles in orangutans are clearly needed.
- Usual innervation: Data are not available.

Hyoglossus (Fig. 10)
- Usual attachments: The hyoglossus is usually differentiated into a **ceratoglossus** and a **chondroglossus** as is usually the case in modern humans (see e.g., Terminologia Anatomica 1998). The ceratoglossus connects the greater horn of the hyoid bone to the tongue, while the chondroglossus mainly connects the body of the hyoid bone (the inferior horn is usually poorly developed or absent) to the tongue.
- Usual innervation: Data are not available.

Styloglossus
- Usual attachments: From stylohyoid process to tongue (e.g., Dean 1984, our dissections). We could not discern, in the orangutans dissected by us, if the **palatoglossus** is present as a separate muscle in these apes, although it should in theory be because the muscle in present in all other apes and many other primates.
- Usual innervation: Data are not available.

Sternohyoideus (VU PP1 LSB = 19.4 g)
- Usual attachments: From the sternum and adjacent regions (e.g. medial end of the clavicles: e.g., Sonntag 1924a) to the hyoid bone; the muscle usually contacts, or lies just next to, its counterpart for most of its length (e.g., Fick 1895a,b, Sonntag 1924a).
- Usual innervation: Data are not available.

Omohyoideus (Fig. 26)
- Usual attachments: Usually has a single head, from the scapula to the hyoid bone.
- Usual innervation: C2 and C3 (Kallner 1956).

- Notes: Contrary to modern humans, in orangutans the omohyoideus usually has no well-defined intermediate tendon separating a superior head and an inferior head. In fact, in *Pongo* the tendon was described by Michaëlis (1903) and Primrose (1899, 1900), but was reported to be absent by most authors, including Sandifort (1840), Fick (1895a,b), Sonntag (1924a,b), Brandes (1932) and Kallner (1956), and in the review done by Ashton & Oxnard (1963) they stated that the tendon is missing in about 71.4% of orangutans. In a few orangutan specimens the omohyoideus is completely missing (e.g., in a specimen reported by Bischoff 1870).
- Synonymy: Coracohyoideus (Tyson 1699), and the omohyoideus probably corresponds to the omo-hyoïdien plus cléido-hyoïdien *sensu* Gratiolet & Alix (1866).

Sternothyroideus (Fig. 9)
- Usual attachments: From the sternum and adjacent regions to the thyroid cartilage.
- Usual innervation: Data are not available.

Thyrohyoideus (Figs. 9–10)
- Usual attachments: From the thyroid cartilage to the hyoid bone.
- Usual innervation: Data are not available.
- Notes: The **levator glandulae thyroideae** muscle, which is present as an anomaly in modern humans, is seemingly not usually present in orangutans.

3.5 Extra-ocular musculature

Muscles of eye (Fig. 7)
- Usual attachments: To our knowledge, there are no detailed published descriptions of these muscles in orangutans. The **rectus inferior, rectus superior, rectus medialis, rectus lateralis, obliquus superior, obliquus inferior, orbitalis** and **levator palpebrae superioris** are consistently found in modern humans and at least some other primate and non-primate mammals, so these muscles are very likely also present in orangutans. Detailed studies of the eye and the extraocular muscles in these and other apes are clearly needed.
- Usual innervation: Data are not available.

CHAPTER 4

Pectoral and Upper Limb Musculature

Serratus anterior (Figs. 14, 21–23, 25–28, 35, 37)
- Usual attachments: From the medial side of the scapula, being well-separated from the levator scapulae, to the ribs (e.g., ribs 1–10: Hepburn 1892, Schück 1913b; 1–12: Beddard 1893; ribs 1–11: Primrose 1899, 1900, Sullivan & Osgood 1927, Stewart 1936).
- Usual innervation: Long thoracic nerve, from C5 and C6 (Hepburn 1892); C6, C7 and C8 (Fick 1895a,b); C5 and C6 (Kohlbrügge 1897); C5, C6 and C7 (Kallner 1956); long thoracic nerve, from C6, C7 and C8 (Schück 1913b); C4–C7 (Sonntag 1924a).
- Function: Sullivan & Osgood (1927) stated that the muscle is a lateral rotator of the scapula, while Tuttle & Basmajian (1976) stated that the caudal part of the serratus anterior exhibited EMG activity during arm-raising in *Pongo*.
- Synonymy: Serratus magnus (Hepburn 1892, Beddard 1893, Primrose 1899, 1900, Sonntag 1924a); serratus anticus (Huntington 1903); pars caudalis of serratus anticus (Schück 1913b); serratus lateralis (Kallner 1956); serratus ventralis thoracis (Jouffroy 1971).

Rhomboideus (VU PP1 LSB rhomboideus + rhomboideus occipitalis = 126.0 g; Figs. 13, 36–37)
- Usual attachments: From the medial border of the scapula to the cervical and thoracic vertebrae.
- Usual innervation: C5 (Hepburn 1892); C4 and C5 (Schück 1913b); C4 and C5 (Kallner 1956).
- Function: Tuttle & Basmajian (1978a,b) stated that the rhomboideus exhibited EMG activity during arm-raising in *Pongo*; low or nil activity was exhibited during crutch-walking and hoisting behavior and pendent suspension in this taxon.
- Notes: The **rhomboideus major**, **rhomboideus minor** are usually not present as distinct entities orangutans; the rhomboideus of orangutans probably corresponds to the rhomboideus major plus rhomboideus minor of modern

humans, the rhomboideus occipitalis being missing in both modern humans and chimpanzees (e.g., Diogo & Wood 2011, 2012).

- Synonymy: Part of rhomboidei (Owen 1830-1831) and of rhomboideus (Sonntag 1924a); part of rhomboidei major and minor (Church 1861-1862); rhomboideus major plus rhomboideus minor (Beddard 1893); spinal part of rhomboideus (Primrose 1899, 1900); rhomboideus major (Kallner 1956).

Rhomboideus occipitalis (VU PP1 LSB rhomboideus + rhomboideus occipitalis = 126.0 g; Figs. 30, 36–37)
- Usual attachments: From the medial border of the scapula to the cranium.
- Usual innervation: C4 and C5 (Schück 1913b, Kallner 1956).
- Notes: See rhomboideus, above.
- Synonymy: Part of rhomboidei (Owen 1830-1831); occipitoscapularis (Wood 1867a, Macalister 1875, Aziz 1981); occipital part of rhomboideus (Primrose 1899, 1900); rhomboideus capitis (Sonntag 1924a); omo-occipitalis (Sullivan & Osgood 1927); rhomboideus minor (Kallner 1956); rhomboideus capitis, levator scapulae minor vel posterior or levator anguli scapulae minor (Jouffroy 1971).

Levator scapulae (VU PP1 LSB = 58.0 g; Figs. 8, 13, 15–16, 36–37)
- Usual attachments: From the cervical vertebrae to the anterior portion of the medial side of the scapula.
- Usual innervation: C3 and C4 (Schück 1913b); Kallner (1956) stated that in the female *Pongo pygmaeus* specimen dissected by her the levator scapulae was innervated by C4–C5 (on one side of the body) and by C3-C4 (on the other side of the body).
- Notes: In *Pongo* Church (1861-1862) described an origin from C1 and Duvernoy (1855-1856) and Primrose (1899, 1900) from C1–C3, while Hepburn (1892), Schück (1913b), Sonntag (1924a) and Sullivan & Osgood (1927), as well as ourselves, found an origin from C1–C4, and Stewart (1936) reported an origin from C1–C5; interestingly, Fick (1895a,b) stated that the origin of the muscle extended to C7.
- Synonymy: Traceloscapularis (Church 1861-1862); levator anguli scapulae (Hepburn 1892, Beddard 1893, Primrose 1899, 1900, Sonntag 1924a); levator scapulae plus pars cranialis of serratus anticus (Schück 1913b); serratus ventralis cervicis, serratus colli, thoracico-scapularis, levator scapulae superioris or rhomboideus profundus (Jouffroy 1971).

Levator claviculae (Figs. 5, 26, 37)
- Usual attachments: From the atlas to the clavicle, passing deep to (covered either laterally or dorsally by) the trapezius.
- Usual innervation: C3 (Schück 1913b, Kallner 1956).
- Notes: Andrews & Groves (1976) stated that in *Pongo* the levator claviculae is mainly lateral to the trapezius, but Shück's (1913a,b), Sullivan & Osgood's (1927), Stewart's (1936), Kallner's (1956) and Ashton & Oxnard's (1963) descriptions and our observations indicate that it is in fact usually deep to the

trapezius. In *Pongo*, according to Primrose (1899, 1900), the insertion of the levator claviculae is at the junction of the middle and lateral third of the bone, according to Sullivan & Osgood (1927) is lateral to the middle of the clavicle, and according to Schück (1913a,b), Michaëlis (1903) and Kallner (1956) is onto the pars acromialis of the clavicle. The **atlantoscapularis posticus** (see Diogo et al. 2009a, Diogo & Wood 2012) is usually not present as a distinct muscle in orangutans.

- Synonymy: Claviotrachélien or acromiotrachélien (Church 1861-1862); omocervicalis, cleido-cervicalis, acromio-cervicalis, levator anticus scapulae (Barnard 1875, Primrose 1899, 1900, Schück 1913b, Kallner 1956); omocervicalis or omotrachelien (Sullivan & Osgood 1927); atlantoscapularis anterior (Ashton & Oxnard 1963); trachelo-acromialis, omotransversarius, omoatlanticus, cleidoatlanticus, levator scapulae major vel anterior, levator scapulae anticus, levator cinguli, or acromio-basilar (Jouffroy 1971).

Subclavius (VU PP1 LSB = 4.5 g; VU PP2 LSB = 8.4 g; Fig. 28)

- Usual attachments: From the first rib, and often also from the second rib, to the clavicle.
- Usual innervation: Data are not available.
- Notes: Hepburn (1892) and Sonntag (1924b) stated that in *Pongo* the subclavius usually originates from ribs 1 and 2, and this was followed by Gibbs (1999) and Gibbs et al. (2002), but in the specimens dissected by Primrose (1899, 1900), Sullivan & Osgood (1927), Stewart (1936) and Kallner (1956), the origin is from rib 1 only. The **costocoracoideus** is usually not present as a distinct muscle in orangutans, but these and most other primates do usually have a ligamentum costocoracoideum, which corresponds to the costocoracoideus muscle of primitive mammals such as monotremes (see Diogo et al. 2009a, Diogo & Wood 2012). It should be noted that Michaelis (1903) described a muscle 'sternoclavicularis' in one orangutan specimen dissected by him, running from the sternal extremity of the clavicle and/or the clavicular extremity of the sternum to the clavicle. However, Kallner (1956) stated that she did not found this muscle in any of the *Pongo* specimens dissected by her or in the other literature, and also suggested that this muscle does not correspond to the costocoracoid ligament, which is often present in this *Pongo*.

Pectoralis major (VU PP1 LSB = 302.0 g; VU PP2 LSB = 316.1 g; Figs. 8, 19, 21–23, 25–27, 37)

- Usual attachments: The pars clavicularis usually runs from the sternum to the proximal humerus; the pars sternocostalis runs from the sternum and ribs to the proximal humerus, inserting proximal to the insertion of the pars clavicularis; the pars abdominalis runs from the ribs and aponeurosis of the external oblique, deep to the pars sternocostalis and pars clavicularis, to the proximal humerus.

- Usual innervation: Medial and lateral pectoral nerves (Hepburn 1892); C5–C8 and T1 (Kallner 1956).
- Function: Tuttle & Basmajian (1976, 1978b) stated that the pars sternocostalis of the pectoralis major exhibited moderate or high EMG activity during descent onto hand and crutch walking in *Pongo*; contrary to *Gorilla* and *Pan*, pars sternocostalis of pectoralis major exhibited very low EMG activity or silence during hoisting behavior in *Pongo*.
- Notes: Some authors (e.g., Church 1861-1862; Sonntag 1924a, Ashton & Oxnard 1963, Andrews & Groves 1976) described a very small clavicular origin of the pectoralis major in orangutans, but most authors (e.g., Chapman 1880, Hartmann 1886, Hepburn 1892, Beddard 1893, Fick 1895a,b, Primrose 1899, 1900, Michaëlis 1903, Sullivan & Osgood 1927, Stewart 1936, Kallner 1956, Stern et al. 1980a) agree that in orangutans there is usually no clavicular origin of the pectoralis major, and this was corroborated by our dissections of *Pongo*. The '**tensor semi-vaginae articulationis humero-scapularis**' *sensu* Macalister (1871) and Huntington (1903), which is also designated as '**sterno-humeralis**', '**sterno-chondro-humeralis**' or '**pectoralis minimus**' by these authors, is a muscle that is occasionally present in modern humans. The '**tensor semi-vaginae articulationis humero-scapularis**' (which is also designated as '**sterno-humeralis**', '**sterno-chondro-humeralis**' or '**pectoralis minimus**'), as well as the '**muscle chondroepitrochlearis**', the **sternalis**, the '**pectoralis quartus**' and the **panniculus carnosus** are usually not present as distinct muscles in orangutans.
- Synonymy: Pectoralis major plus lower portion of pectoralis minor (Hartmann 1886); pectoralis major plus second half of pectoralis minor (Beddard 1893); pectoralis major + pectoralis abdominalis (Huntington 1903); pectoralis major + pectoralis quartus (Preuschoft 1965); pectoralis major + pectoralis abdominis, abdominalis, and/or chondroepitrochlearis quartus (Richmond 1993, Gibbs 1999).

Pectoralis minor (VU PP1 LSB = 26.0 g; VU PP2 LSB = 55.2 g; Figs. 6, 28)
- Usual attachments: From the proximal humerus and/or the glenohumeral joint capsule and from the coracoid process of the scapula, to the ribs (e.g., ribs 3–4: Beddard 1893, Primrose 1899, 1900, Stewart 1936; ribs 2–5: Sonntag 1924a; ribs 3–5: Sullivan & Osgood 1927).
- Usual innervation: Medial pectoral nerve (Hepburn 1892).
- Notes: In the orangutans dissected by us there was an insertion onto the coracoid process, and such an insertion was also reported in orangutans by Hartmann (1886), Hepburn (1892), Fick (1895a,b), Kohlbrügge (1897), Sonntag (1924a), Sullivan & Osgood (1927) and Kallner (1956). The **pectoralis tertius** ('**xiphihumeralis**') (see Diogo et al. 2009a, Diogo & Wood 2012) is usually not present as a distinct muscle in orangutans.
- Synonymy: Upper portion of pectoralis minor (Hartmann 1886); first half of pectoralis minor (Beddard 1893).

Infraspinatus (VU PP1 LSB = 137.0 g; VU PP2 LSB = 173.1 g; Figs. 35–36, 38)
- Usual attachments: From the infraspinous fossa of the scapula, the scapular spine and the infraspinatus fascia to the greater tuberosity of the humerus and in at least some cases also to the capsule of the glenohumeral joint.
- Usual innervation: Suprascapular nerve, from C5 and C6 (Hepburn 1892); Suprascapular nerve (Kallner 1956).
- Function: Tuttle & Basmajian (1978a) stated that the infraspinatus exhibited nil EMG activity during hoisting behavior in *Pongo*.

Supraspinatus (VU PP1 LSB = 71.0 g; VU PP2 LSB = 97.4 g; Figs. 30, 36, 38)
- Usual attachments: From the supraspinous fossa of the scapula, the supraspinatus fascia and often also from the scapular spine and/or acromion, mainly to the greater tuberosity of the humerus and in at least some cases to the capsule of the glenohumeral joint; the muscle is sometimes fused with the infraspinatus (Gibbs 1999).
- Usual innervation: Suprascapular nerve, from C5 and C6 (Hepburn 1892); Suprascapular nerve (Kallner 1956).
- Function: Tuttle & Basmajian (1978a) stated that the supraspinatus exhibited moderate or high EMG activity during arm-raising, quiet tripedal stance and descent onto hand in *Pongo*; low or nil activity was exhibited during hoisting behavior in *Pongo*.
- Notes: The muscle **scapuloclavicularis**, occasionally present in modern humans, has not been described in orangutans nor was it found in the orangutans dissected by us. Potau et al. (2009) studied the relative masses of the deltoideus, subscapularis, supraspinatus, infraspinatus and teres minor in chimpanzees, orangutans and modern humans and showed that for all of the proportional values in *Pongo pygmaeus* there is marked overlap between the ranges of values in the two species. According to Potau et al. (2009), this may indicate that the functional requirements of the glenohumeral joint are similar in a great ape with fundamentally suspension/vertical climbing locomotion and in a bipedal great ape that uses the upper extremity for essentially manipulative functions.

Deltoideus (VU PP1 LSB = 300.0 g; VU PP2 LSB = 422.3 g; Figs. 6, 25–29, 35–38)
- Usual attachments: Mainly from the lateral portion of the clavicle (pars clavicularis: e.g., lateral 1/2 of clavicle according to Barnard 1875, Sonntag 1924a, Sullivan & Osgood 1927 and Kallner 1956; lateral 1/3 of clavicle according to Primrose 1899, 1900 and Ashton & Oxnard 1963; lateral 1/4 of clavicle according to Andrews & Groves 1976), the acromion (pars acromialis) and the spine of the scapula and often the infraspinous fascia (pars spinalis) to the humerus.
- Usual innervation: Axillary nerve (Hepburn 1892); Axially nerve from C4, C5 and C6 (Kohlbrügge 1897); axially nerve from C5 and C6 (Kallner 1956).

- Function: Tuttle & Basmajian (1978a,b) stated that the deltoideus exhibited moderate to high EMG activity during arm-raising, descent onto hand and hoisting behavior in *Pongo*.

Teres minor (VU PP1 LSB = 29.0 g; VU PP2 LSB = 40.1 g; Figs. 30, 38)
- Usual attachments: From the infraspinatus fascia and the lateral border of the scapula to the greater tuberosity of the humerus (usually the muscle does not extend to the humeral shaft, distally to this tuberosity).
- Usual innervation: Axillary nerve (Hepburn 1892, Kohlbrügge 1897, Kallner 1956).
- Function: Tuttle & Basmajian (1978a) stated that the teres minor exhibited low to nil EMG activity during arm-raising (a striking contrast with the activity pattern of this muscle during arm-raising in humans) and hoisting behavior in *Pongo*.

Subscapularis (VU PP1 LSB = 172.0 g; VU PP2 LSB = 230.3 g; Figs. 28, 30, 37, 39)
- Usual attachments: From the subscapular fossa of the scapula to the lesser tuberosity of the humerus, and in at least some case also to the humeral shaft just distal to this tuberosity (e.g., Beddard 1893; our specimen HU PP1).
- Usual innervation: Subscapular nerves (Hepburn 1892, Kohlbrügge 1897, Kallner 1956).
- Function: Tuttle & Basmajian (1978a) stated that the subscapularis exhibited EMG activity during arm-raising in *Pongo*; low or nil activity was exhibited during hoisting behavior in *Pongo*.
- Notes: In a *Pongo* specimen dissected by Sullivan & Osgood (1927) the subscapularis had a small 'upper' (anterior *sensu* the present work) bundle that was quite independent of the rest of the muscle.

Teres major (VU PP1 LSB = 102.0 g; VU PP2 LSB = 149.7 g; Figs. 12, 28–30, 35–38)
- Usual attachments: From the lateral border and inferior angle of the scapula to the proximal portion of the intertubercular groove of the humerus by means of a distal tendon that is usually fused to the distal tendon of the latissimus dorsi.
- Usual innervation: Subscapular nerves (Hepburn 1892, Kohlbrügge 1897, Kallner 1956).
- Function: Tuttle & Basmajian (1976) stated that the teres major exhibited EMG activity during hoisting behavior in *Pongo*.
- Notes: The distal tendons of the latissimus dorsi and of the teres major are usually partially or completely fused to each other in *Pongo* (corroborated by, e.g., Hepburn 1892, Church 1861-1862, Barnard 1875, Beddard 1893, Primrose 1899, 1900, Sonntag 1924a, Sullivan & Osgood 1927, Kallner 1956, Ashton & Oxnard 1963, Miller 1932, Stewart 1936, one of the few exceptions found in the recent literature concerns the work of Payne 2001, who suggested that the distal tendons of the two muscles were not fused in the three orangutans

dissected by her; the fusion between the tendons was also found in one adult *Pongo* specimen recently dissected by us, but the tendons were partially separated in a orangutan neonate recently dissected by S. Dunlap: pers. comm.).

Latissimus dorsi (VU PP1 LSB = 637.0 g; Figs. 12–14, 25–30, 35–36, 38)
- Usual attachments: Usually from the vertebrae, ribs (but costal origin may be missing), thoracolumbar fascia and, often, directly and/or indirectly from the pelvis, to the intertubercular groove of the humerus.
- Usual innervation: Thoracodorsal nerve (Hepburn 1892, Kallner 1956).
- Function: Tuttle & Basmajian (1976, 1978b) stated that the latissimus dorsi exhibited moderate to high EMG activity during hoisting behavior and crutch-walking in *Pongo*.
- Notes: Numerous authors have reported the marked subdivision of the latissimus dorsi into at least 2 bundles in *Pongo*. For instance, Church (1861-1862) reported a specimen in which the portion of the latissimus dorsi that originated from the thoracic vertebrae forms a distinct bundle that is separated from the rest of the muscle by a septum of dense tissue and that inserts onto the external fascia of the arm and the humerus, together with the teres major; the rest of the muscle is inserted an inch and a half below the intertubercular groove of the humerus. Beddard (1893) described a specimen in which the latissimus dorsi was medially divided into two parts. Primrose (1899, 1900) described a specimen with a small muscular slip of the muscle passing on a plane dorsal to the main part of the muscle and to the dorsoepitrochlearis to insert to the humerus together with the tendon of the teres major. Schück (1913a) reported a specimen in which the latissimus dorsi had a main body and a well-separated anterior head. Stewart (1936) reported a specimen in which, on the left side of the body, there was an additional slip arising from rib 10 and appearing to be continuous with the external oblique; a small cranial portion, representing about one-fourth of the origin, fused with the teres major and inserted onto the humerus, while the larger portion of the latissimus dorsi fused with the dorsoepitrochlearis and then formed a tendon that inserted onto the humerus. Lastly, Kallner's (1956) Fig. 1 seems to suggest that the latissimus dorsi has a smaller anterior head and a broader posterior, main body, while Payne (2001) reported three specimens in which the latissimus dorsi had two distinct bellies, the more proximal inserting onto the proximo-medial humeral shaft above the teres major and giving origin to the tendon of the dorsoepitrochlearis, while the distal belly inserted below the teres major.

Dorsoepitrochlearis (VU PP2 LSB = 31.2 g; Figs. 27–28, 30, 38)
- Usual attachments: Mainly from the distal portion of the latissimus dorsi to the medial epicondyle of the humerus and occasionally also to the adjacent supracondylar ridge of the humerus; only in a few cases the dorsoepitrochlearis attachment extends distally to attach directly onto the olecranon process of the ulna.

- Usual innervation: Radial nerve (Hepburn 1892, Sullivan & Osgood 1927, Kallner 1956, Kawashima et al. 2007).
- Notes: As noted by Aiello & Dean (1990), in non-human hominoids the dorsoepitrochlearis is usually mainly attached onto the medial epicondyle, the intermuscular septum and/or other surrounding structures, but not onto the olecranon process or the olecranon fascia. Regarding *Pongo*, Barnard (1875) and Fick (1895a,b) found an insertion onto the medial epicondyle and intermuscular septum, Chapman (1880), Beddard (1893), Primrose (1899), (1900), Michaëlis (1903), Sonntag (1924a), Sullivan & Osgood (1927), Miller (1932), Andrews & Groves (1976) found an insertion onto the medial epicondyle only, as we did, while Schück (1913a) refers to a bony insertion onto the humerus, Hepburn (1892) and Ashton & Oxnard (1963) to the intermuscular septum, and only Church (1861-1861) refers to a bony insertion onto the olecranon process of the ulna.
- Synonymy: Latissimo-condylus or latissimo-epitrochlearis (Barnard 1875); latissimo-condyloideus (Chapman 1880, Hepburn 1892, Primrose 1899, 1900, Sullivan & Osgood 1927, Loth 1931); latissimo-tricipitalis (Schück 1913a, Kallner 1956); tensor fasciae antebrachii, anconeus accessorius, accessorius latissimus dorsi, dorso-antebrachialis, anconeus quintus, anconeus longus, extensor cubiti (Jouffroy 1971).

Triceps brachii (VU PP1 LSB = 297.0 g; VU PP2 LSB = 324.7 g; Figs. 27–30, 35, 38–39)
- Usual attachments: From at least half of the length of the lateral border of the scapula (caput longum) and from the shaft of the humerus (caput laterale and caput mediale) to the olecranon process of the ulna.
- Usual innervation: Radial nerve (Hepburn 1892, Sonntag 1924a, Kallner 1956).
- Notes: In *Pongo* the long head of the triceps brachii usually originates from half or more than half (1/2) of this border: e.g., Sonntag (1924a), Sullivan & Osgood (1927), Loth (1931), and Gibbs (1999) refer to 1/2, we also found 1/2, and Primrose (1899, 1900) refers to 2/3. It should be noted that the triceps brachii of modern humans includes the **articularis cubiti**, which is listed in Terminologia Anatomica (1998) as a muscle that is usually present in modern humans, but this term refers to a subdivision of the triceps brachii that runs from the main body of that muscle to the posterior aspect of the capsule of the elbow joint, thus lifting the capsule away from the joint; it should not to be confused with the muscle anconeus (see below). To our knowledge, the articularis cubiti has not been found in orangutans.
- Synonymy: Multiceps extensor cubiti (Barnard 1875); triceps extensor cubiti (Hepburn 1892); triceps (Sullivan & Osgood 1927).

Brachialis (VU PP2 LSB = 248.3 g; Figs. 29–31)
- Usual attachments: From the humeral shaft (i.e., distal to the surgical neck of the humerus) to the ulnar tuberosity.

- Usual innervation: Musculocutaneous nerve (Hepburn 1892, Sonntag 1924a, Kallner 1956, Koizumi & Sakai 1995).
- Notes: According to Gibbs (1999), in 1/5 *Pongo* the origin of the brachialis is split into two parts from the anteromesial and anterolateral surfaces of the humerus respectively, which then fuse distally. For instance, Sonntag (1924a) reported a specimen in which the brachialis had two parts, which only came together in the distal half. The mesial part arised from the median half of the front of the humerus up to a point about a sixth of an inch above the lowest point of the deltoid insertion, and was fused with the coracobrachialis; the lateral part arised from the lateral half of the humerus and extended well above the insertion of deltoideus, the two parts joining each other distally. Sullivan & Osgood (1927) described a specimen in which the brachialis seemed to consist of three distal parts: a central part made up of long bundle from the middle third of the humerus, a medial segment consisting of slightly shorter bundles from the epicondylar line and shaft of the humerus, and a lateral portion made up of still shorter bundles coming from the lateral epicondylar line. Also, Payne (2001) reported three specimens in which the brachialis had two bellics, the proximal one arising by a tendon and the distal one arising from bone, from the anterior aspect of the humeral shaft; these two bellies fused onto a common tendon and inserted onto the ulna.
- Synonymy: Brachialis anticus (Hepburn 1892, Beddard 1893, Primrose 1899, 1900, Sonntag 1924a).

Biceps brachii (VU PP1 LSB = 165.0 g; VU PP2 LSB = 224.9 g; Figs. 31, 39, 26–30)
- Usual attachments: From the proximal humerus (caput breve) and the supraglenoid tubercle of the scapula (caput longum), to the bicipital tubercle of the radius (common tendon).
- Usual innervation: Musculocutaneous nerve (Hepburn 1892, Sonntag 1924a, Kallner 1956, Koizumi & Sakai 1995).
- Function: Tuttle & Basmajian (1992) stated that contrary to *Gorilla* and *Pan*, which exhibited low or nil activity of the biceps brachii in supination, EMG in *Pongo* showed that this muscle was consistently active during supination.
- Notes: In hominoids such as *Hylobates*, *Gorilla*, *Pan* and modern humans, the biceps brachii is usually prolonged distally by a bicipital aponeurosis ('lacertus fibrosus' in *Gorilla*, *Pan* and modern humans; often 'lacertus carnosus' in *Hylobates*), which is commonly associated with the fascia covering forearm muscles such as the pronator teres. In orangutans the bicipital aponeurosis is usually missing, e.g., most specimens, including those described by Fick (1895a,b), Sonntag (1924a), Sullivan & Osgood (1927) and Kallner (1956), and dissected by us, lack this structure, with exception of a few specimens, such as the orangutan described by Primrose (1899, 1900).
- Synonymy: Biceps flexor cubiti (Hepburn 1892).

Coracobrachialis (VU PP1 LSB = 31.0 g; VU PP2 LSB = 46.6 g; Figs. 28, 30, 39)
- Usual attachments: From the coracoid process of the scapula to the proximal portion of the humerus, being usually (e.g., Chapman 1880; Primrose 1899, 1900, Sonntag 1924a, Sullivan & Osgood 1927; our specimen HU PP1) pierced by the musculocutaneous nerve.
- Usual innervation: Musculocutaneous nerve (Hepburn 1892, Primrose 1899, 1900, Sonntag 1924a, Kallner 1956, Koizumi & Sakai 1995); in some *Pongo* specimens innervation is from the median nerve (Kohlbrügge 1897, Howell & Straus 1932); musculocutaneous nerve, but also by a superficial and a deep branch arising from the medial and posterior cords of the brachial plexus (Kawashima et al. 2007).
- Notes: The observations of most authors, as well as our dissections, show that the usual condition for orangutans is that the **coracobrachialis superficialis/ longus** and **coracobrachialis profundus/coracobrachialis brevis** are not present as distinct structures, i.e., the coracobrachialis has a single bundle that corresponds to the **coracobrachialis medius/coracobrachialis proprius** of other mammals. However, the coracobrachialis profundus was reported in one *Pongo* specimen described by Kallner (1956) (a distinct 'coracobrachialis longus' —or 'superficialis'—was also reported in one of the *Pongo* specimens dissected by Kallner 1956, but the use of these names is most likely erroneous, because in Kallner's illustrations there is nothing resembling the coracobrachialis longus of other tetrapods).
- Synonymy: Coracobrachiales medius (Parsons 1898b, Jouffroy 1971).

Pronator quadratus (VU PP1 LSB = 9.0 g; VU PP2 LSB = 8.8 g; Figs. 31, 43)
- Usual attachments: From the distal portion of the ulna to the distal portion of radius.
- Usual innervation: Anterior interosseous branches of the medial nerve (Sonntag 1924a); medial nerve (Kallner 1956).
- Function: According to Tuttle (1969), the trend toward reduced development of the pronator muscles in the African apes may be associated with the fact that they do not engage as frequently as orangutans in extreme arboreal movements that require a wide range of supination and pronation of the forearm; furthermore, the African apes, particularly *Gorilla*, require considerable stability at the elbow joint to maintain the fully extended forearm against the compressive forces incurred during knuckle-walking.

Flexor digitorum profundus (VU PP1 LSB = 261.3 g; VU PP2 LSB = 312.7 g; Figs. 31, 41–42)
- Usual attachments: From the radius, ulna, and interosseous membrane to the distal phalanges of digits 2, 3, 4 and 5 and, sometimes, to digit 1 through a tendon that is often thin and/or not continuous to the main body of the muscle (see Notes below).

- Usual innervation: Median (usually to digits 2 and 3, and digit 1 if there is a portion of the muscle corresponding to the flexor pollicis longus of humans) and ulnar (usually to digits 4 and 5) nerves (Hepburn 1892, Primrose 1899, 1900, Sonntag 1924a, Kallner 1956).

- Function: Tuttle (1969) stated that a tendency toward shortening of the 'long digital flexor tendons' (as usually seen in *Pan* and *Gorilla*) was noted in some *Pongo* adults in the Yerkes colony that had 'fist-walked'; however, the few adult animals that had elected a palmigrade mode of progression retained the capacity to fully extend the fingers as well as to dorsiflex the wrist.

- Notes: Among the primates dissected by us, hylobatids and modern humans are the only ones in which the flexor pollicis longus is usually present as a distinct, independent muscle (e.g., the **flexor pollicis longus** is usually not present as a distinct muscle in orangutans: Diogo & Wood 2011, 2012). That is, when some authors state that there is a 'flexor pollicis longus' in a orangutan they are either referring to the part of the flexor digitorum profundus that goes to digit 1, or to the belly of this muscle that often goes to both digits 1 and 2, and not really to a distinct muscular belly going exclusively to digit 1 as that usually found in hylobatids and modern humans (see Synonymy below). In fact, in *Pongo*, *Gorilla* and *Pan* the tendon to digit 1 is effectively often absent or vestigial, as corroborated in the specimens dissected by us and by others. For instance, Straus (1942b) compiled evidence from his own dissections of hominoids and data available on the literature; among 27 orangutans the tendon was entirely absent in 89% (24 of 27), rudimentary and functionless in 7% (2 of 27), and completely developed in but 4% (1 of 27); this tendon was thus eliminated physiologically in 96% of orangutans.

- Synonymy: Flexor profundus plus flexor longus pollicis (Church 1861-1862); flexor digitorum communis profundus (Barnard 1875); flexor profundus digitorum (Hepburn 1892, Beddard 1893, Primrose 1899, 1900, Sonntag 1924a); flexor profundus (Richmond 1993).

Flexor digitorum superficialis (VU PP1 LSB = 143.2 g; VU PP2 LSB = 172.3 g; Figs. 31, 33–34, 44)

- Usual attachments: From the radius (caput radiale), the ulna and the medial epicondyle of the humerus (caput humeroulnare) to the middle phalanges of digits 2–5.

- Usual innervation: Median nerve (Hepburn 1892, Sonntag 1924a, Kallner 1956).

- Function: According to Tuttle (1969) in great apes the fasciculi of the flexor digitorum superficialis and/or flexor digitorum profundus to the individual digits, especially digit 2, frequently appear as entities in the middle of the forearm and may be traced proximally at least to this level by following the individual tendons of the flexor digitorum superficialis/profundus muscles. According to Tuttle this is probably related to the ability to flex the fingers independently. This is of considerable advantage to orangutans and chimpanzees in that it allows the animals to hold a number of small twigs with some digits while the remaining fingers reach for additional supports. If some of the twigs break, the individual does not have to open the whole hand in order to grasp replacements.

- Notes: Hepburn (1892), Kohlbrügge (1897), Primrose (1899, 1900), Sonntag (1924a), Loth (1931), Jouffroy (1971, Gibbs (1999) and Payne (2001) refer to an origin of the flexor digitorum superficialis of *Pongo* from the ulna, radius and the medial epicondyle of the humerus and/or the common flexor tendon, as we found in our dissections, and only a few authors (e.g., Beddard 1893, Kallner 1956) stated that there is no ulnar origin, while only Michaëlis (1903) did not refer to an origin from the radius.
- Synonymy: Flexor digitorum sublimis (Sullivan & Osgood 1927); flexor sublimis digitorum (Church 1861-1862, Hepburn 1892, Beddard 1893, Primrose 1899, 1900, Sonntag 1924a); flexor secundi internodii digitorum (Jouffroy 1971).

Palmaris longus (VU PP1 LSB = 23.1 g; VU PP2 LSB = 33.4 g; Figs. 31, 40)
- Usual attachments: Mainly from medial epicondyle of the humerus to the palmar aponeurosis.
- Usual innervation: Median nerve (Hepburn 1892, Sonntag 1924a, Kallner 1956).
- Notes: In the *Pongo* specimens described by Church (1861-1862; 1 specimen), Chapman (1880; 1 specimen), Hepburn (1892; 1 specimen), Beddard (1893; 1 specimen), Fick (1895a,b; 1 specimen), Primrose (1899, 1900; 1 specimen), Michaëlis (1903; 1 specimen), Sonntag (1924a; 1 specimen), Kallner (1956; 2 specimens), Payne (2001; 3 specimens), Oishi et al. (2008, 2009; 3 specimens) and dissected by us, the palmaris longus was always present, and according to Loth (1931), Gibbs (1999) and Gibbs et al. (2002) it is present in all of the specimens of this genus, although Traill (1821) suggested that it was missing on one side of a orangutan dissected by him.

Flexor carpi ulnaris (VU PP1 LSB = 36.4 g; VU PP2 LSB = 69.2 g; Figs. 31, 40–42)
- Usual attachments: Usually from the ulna (caput ulnare) and the medial epicondyle of the humerus (caput humerale) to the pisiform.
- Usual innervation:
- Notes: A bony origin of the flexor carpi ulnaris from the ulna and humerus was found in the *Pongo* specimens dissected by us and described by Hepburn (1892), Beddard (1893), Primrose (1899, 1900), Sonntag (1924a), Sullivan & Osgood (1927) and Kallner (1956). According to the review of Gibbs (1999) the ulnar nerve passes between the two heads in all apes, with the exception of 1/5 *Pongo* in which it runs along the deep surface of the muscle, as described by Sullivan & Osgood (1927). The structure that Kallner (1956) designated as 'anconeus sextus' clearly seems to correspond to the **epitrochleoanconeus**, and not to the anconeus, *sensu* the present study, as the name 'anconeus sextus' has often been used by other authors to designate the epitrochleoanconeus, and also because Kallner (1956) stated

that the 'anconeus sextus' was innervated by the ulnar nerve. However, confusingly, in her table 21 Kallner (1956) listed the insertions of both the anconeus and of the epitrochleoanconeus *sensu* the present study, and the only three cases where she seems to be actually referring to the epitrochleoanconeus in this table are in the lines about Kolhbrügge's 1897, Testut's 1884 and her own studies. In the two *Pongo pygmaeus* specimens dissected by Kallner (1956) this 'anconeus sextus' connects the distal portion of the humerus and the intermuscular septum to the olecranon process of the ulna. Be that as it may, an epitrochleoanconeus was seemingly only reported in *Pongo* by these three authors (Testut 1884; Kohlbrügge 1897; Kallner 1956), and was not found in any orangutan specimen dissected by us and by most authors, including Church (1861-1862), Hepburn (1892), Beddard (1893), Fick (1895a,b), Primrose (1899, 1900), Michaëlis (1903), Sonntag (1924a), Sullivan & Osgood (1927) and Oishi et al. (2008), so the muscle clearly seems to be usually missing in this genus.

Flexor carpi radialis (VU PP1 LSB = 46.8 g; VU PP2 LSB = 78.5 g; Figs. 31-40)
- Usual attachments: From the medial epicondyle of the humerus and usually also from the radius, to the base of metacarpal II and often also to the base of metacarpal III.
- Usual innervation: Median nerve (Hepburn 1892, Sonntag 1924a, Kallner 1956).
- Notes: In *Pongo*, an exclusive insertion of the flexor carpi radialis onto metacarpal II was described by Beddard (1893; 1 specimen), Kohlbrügge (1897; 1 specimen), Primrose (1899, 1900; 1 specimen) and Sullivan & Osgood (1927; 1 specimen). However, one cannot be sure if these descriptions might have been influenced by the *a priori* expectations of the authors, because in the literature it was often said that in apes the muscle was exclusively onto this bone. In fact, in the *Pongo* specimens described by Hepburn (1892; 1 specimen), and Sonntag (1924a; 1 specimen), as well as in at least two specimens dissected by us (GWU PP1 and VU PP1), the muscle actually goes to both metacarpals II and III (Kallner 1956 stated that in the two specimens dissected by her the insertion was onto metacarpal I instead). In the survey of the literature by Gibbs (1999), she stated that an insertion onto metacarpals II and III was found in 3 out of 7 orangutans, but the fact that we found this condition in two orangutans dissected by us, and also that even if we would take into account Gibbs' numbers we would now have, with our two specimens, a total of 5 out of 9 orangutans with a double insertion, seem to suggest that a double insertion is effectively often found in *Pongo*. In *Pongo* the flexor carpi radialis usually has bony origins from at least the humerus and the radius, as found in the *Pongo* specimens reported by Hepburn (1892), Beddard (1893), Fick (1895a,b), Kohlbrügge (1897), Primrose (1899, 1900), Sonntag (1924a), Sullivan & Osgood (1927), Kallner (1956) and dissected by us.
- Synonymy: Palmaris major or flexor manus radialis (Jouffroy 1971).

Pronator teres (VU PP1 LSB = 41.3 g; VU PP2 LSB = 64.6 g; Figs. 31, 41)
- Usual attachments: From the humerus (caput humerale) and usually also from the ulna (caput ulnare), to the radius.
- Usual innervation: Median nerve, which passes between the ulnar and humeral heads of the muscle (Chapman 1880, Hepburn 1892, Sonntag 1924a, Sullivan & Osgood 1927, Kallner 1956).
- Notes: The *Pongo* specimens reported by Chapman (1880), Hepburn (1892), Beddard (1893), Fick (1895a,b), Primrose (1899, 1900), Sonntag (1924a,b), Kallner (1956), Oishi et al. (2008, 2009) and dissected by us have an ulnar head and a humeral head separated by the median nerve, although Lewis (1989) said that a distinct ulnar head is not present in apes. According to Parsons (1898b), Loth (1931) and Jouffroy (1971) the two heads are present in 100%, 73% and 70% of orangutans, respectively, although Gibbs (1999) confusingly (and seemingly erroneously) stated that the two heads are found only in 3/7 orangutans.
- Synonymy: Pronator radii teres (Chapman 1880, Hepburn 1892, Beddard 1893, Primrose 1899, 1900, Sonntag 1924a); pronator radii profundus or epitrochleoradialis (Jouffroy 1971).

Palmaris brevis (VU PP2 LSB = 0.6 g; Fig. 33)
- Usual attachments: From the pisiform and/or the flexor retinaculum to the skin of the medial border of the palm; in some orangutans the muscle is missing (see Notes below).
- Usual innervation: Data are not available.
- Notes: The palmaris brevis muscle was not present in the *Pongo* specimen dissected by Hepburn (1892) and the two specimens dissected by Kallner (1956), but Loth (1931) stated that it sometimes present in this taxon and it was present in the three specimens where we analysed this feature in detail (VU PP1, VU PP2, GWU PP1), so according to our own review of the literature and of the data obtained from our own dissections, it is present in 3/6, i.e., in 50% of orangutans. The **flexor digitorum brevis manus** and **palmaris superficialis** (e.g., Diogo et al. 2009a, Diogo & Wood 2012) are usually not present as distinct muscles in orangutans.

Lumbricales (VU PP1 LSB lumbricalis 1 = 2.5 g, lumbricalis 2 = 3.4 g, lumbricalis 3 = 1.8 g, lumbricalis 4 = 1.1 g; VU PP2 LSB lumbricalis 1 = 4.8 g, lumbricalis 2 = 5.6 g, lumbricalis 3 = 3.0 g, lumbricalis 4 = 2.5 g; Figs. 33-34, 41–42)
- Usual attachments: From the dorsal surfaces of the tendons of the flexor digitorum profundus to digits 2, 3, 4 and 5, to the radial side of the proximal phalanx and the extensor expansion of digits 2 (lumbricalis 1), 3 (lumbricalis 2), 4 (lumbricalis 3) and 5 (lumbricalis 4).
- Usual innervation: Data are not available.

- Notes: The **intercapitulares** (see, e.g., Jouffroy 1971) are usually not present in orangutans. Regarding the **contrahentes digitorum**, these muscles were missing and/or aponeurotic in the *Pongo* specimens dissected by Church (1861-1862), Brooks (1886a), Hepburn (1892), Primrose (1899, 1900), Sonntag (1924a), Kallner (1956) and the specimens dissected by us, although a fleshy contrahens to digit 2 was possibly present in the right hand of the specimen described by Jouffroy & Lessertisseur (1958) and in one specimen illustrated by Langer (1879), while Hartmann (1886) described contrahentes to digits 4 and 5 in an orangutan. That is, the contrahentes digitorum are seemingly not usually present as distinct, fleshy muscles in orangutans.

Adductor pollicis (VU PP1 LSB = 11.4 g; VU PP2 LSB = 21.2 g; Figs. 33–34, 47, 49)

- Usual attachments: The caput obliquum and caput transversum are usually well-differentiated, connecting mainly the metacarpal III, the contrahens fascia, and often at least some carpal bones and/or ligaments, to the metacarpophalangeal joint, as well as to the proximal phalanx and sometimes also to the distal phalanx of the thumb (directly by means of a thin tendon and/or indirectly by means of a ligament: e.g., Tocheri et al. 2008, Tuttle & Cortright 1988, and our specimens HU PP1 and GWU PP1) and occasionally to metacarpal I (see Notes below).
- Usual innervation: Deep division of the ulnar nerve (Brooks 1887, Hepburn 1892, Kallner 1956).
- Notes: There has been much controversy regarding the homologies of the thenar muscles of primate and non-primate mammals. This subject has been discussed in detail in the recent studies of Diogo et al. (2009a) and particularly of Diogo & Abdala (2010) and Diogo & Wood (2011, 2012) and here we summarize the main conclusions of those studies. The '**interosseous volaris primus of Henle**' of modern human anatomy corresponds to a **thin, deep additional slip of the adductor pollicis (TDAS-AD** *sensu* Diogo & Abdala 2010 and Diogo & Wood 2011; **adductor pollicis accessorius** *sensu* Diogo et al. 2012), and not to the **flexor brevis profundus 2** of 'lower' mammals, as suggested by some authors and as suggested by the use of the name 'interosseous volaris primus of Henle'. Within orangutans, the adductor pollicis accessorius is seemingly always, or almost always, absent. Only Tuttle (1969) stated that the 'interosseous volaris primus of Henle' of modern human anatomy may be found in a few orangutans, but he did not clarified in which *Pongo* specimens was this structure found and/or who dissected these specimens. To our knowledge, this structure was never described in orangutans by other authors, and it was also not found in the orangutans dissected by us. Regarding the '**deep head of the flexor pollicis brevis**' of modern human anatomy, this corresponds to the flexor brevis profundus 2 of 'lower' mammals. This structure is clearly found in the orangutans dissected

by us and in most of the specimens dissected by other authors, who often refer to this structure as the 'deep head of the flexor pollicis brevis'. Regarding the **'superficial head of the flexor pollicis brevis'** of modern human anatomy, this thus seems to correspond to a true **flexor pollicis brevis**. The true flexor pollicis brevis and the **opponens pollicis** likely derive from the **flexor brevis profundus 1** of 'lower' mammals, and they are present as distinct muscles in the vast majority of orangutans (see below). Regarding the attachments of the adductor pollicis, Church (1861-1862), Brooks (1887) and Sonntag (1924a) stated that in the three *Pongo* specimens dissected by them (one specimen each) the adductor pollicis inserted only onto the proximal phalanx of the thumb, but Michaëlis (1903), Duckworth (1904), Kallner (1956) and Jouffroy & Lessertisseur (1960) suggested that a partial insertion onto a small part of metacarpal I is frequently found in this genus, and we did find such a partial insertion onto metacarpal I in the single orangutan specimen dissected by us in which we could discern this feature appropriately (GWU PP1). What seems to be clear is that all the *Pongo* specimens dissected by us and described in the literature do not have an insertion onto most of metacarpal I, i.e., in this respect *Pongo* is similar to the other great apes and to modern humans and is different from *Hylobates* (Diogo & Wood 2011, 2012).

- Synonymy: Adductor obliquus pollicis plus adductor transversus pollicis (Primrose 1899, 1900).

Interossei palmares (VU PP1 LSB interosseus palmaris 1 = 6.6 g, interosseus palmaris 2 = 4.5 g, interosseus palmaris 3 = 3.5 g; VU PP2 LSB interosseus palmaris 1 = 8.2 g, interosseus palmaris 2 = 6.4 g, interosseus palmaris 3 = 4.8 g; Figs. 34, 48–49)

- Usual attachments: Interosseus palmaris 1 runs mainly from metacarpals II to the ulnar side of the proximal phalanx of digit 2; interosseus palmaris 2 runs mainly from metacarpal IV to the radial side of the proximal phalanx of digit 4; interosseus palmaris 3 runs mainly from metacarpal V to the radial side of the proximal phalanx of digit 5. These three muscles thus mainly contribute to adductor these three digits.

- Usual innervation: Deep branch of ulnar nerve (Brooks 1887, Hepburn 1892, Kallner 1956).

- Notes: As explained above (see adductor pollicis and interossei palmares), the **flexor brevis profundus 2** of 'lower' mammals corresponds very likely to the **'deep head of the flexor pollicis brevis'** of modern human anatomy; the **'superficial head of the flexor pollicis brevis'** of modern human anatomy, as well as the **opponens pollicis**, derive very likely from the **flexor brevis profundus 1** of 'lower' mammals, while the **flexor digiti minimi brevis** and the **opponens digiti minimi** derive very likely from the **flexor brevis profundus 10** of 'lower' mammals (for recent reviews, see Diogo et al. 2009a, Diogo & Abdala 2010 and Diogo & Wood 2012). In extant hominoids except *Pan* the **flexores breves profundi** 3, 5, 6 and 8 are fused with the **intermetacarpales** 1, 2, 3 and 4, forming the interossei dorsales 1, 2, 3 and 4, respectively; the **interossei palmares** 1, 2 and 3 of extant hominoids except *Pan* thus correspond respectively to the flexores breves profundi 4, 7 and 9 of 'lower' mammals and of primates such as chimpanzees (e.g., Lewis 1989, Diogo et al. 2009a, and Diogo & Abdala 2010). The **interossei accessorii** are not present as distinct muscles in orangutans.

- Synonymy: Palmar interossei and probably part of abductor pollicis brevis (Church 1861-1862); palmar interossei plus interosseous volaris primus (Brooks 1887, Hepburn 1892); palmar interossei plus deep head of flexor pollicis brevis (Primrose 1899, 1900).

Interossei dorsales (VU PP1 LSB interosseus dorsalis 1 = 22.7 g, interosseus dorsalis 2 = 13.2 g, interosseus dorsalis 3 = 7 g, interosseous dorsalis 4 = 6.9 g; VU PP2 LSB interosseus dorsalis 1 = 28.3 g, interosseus dorsalis 2 = 16.3 g, interosseus dorsalis 3 = 9.5 g, interosseus dorsalis 4 = 9.3 g; Figs. 34–49)

- Usual attachments: As in modern humans, the interosseus dorsalis 1 originates mainly from metacarpals I and II, the interosseus dorsalis 2 mainly from metacarpals II and III, the interosseus dorsalis 3 from metacarpals III and IV, and the interosseus dorsalis 4 from metacarpals IV and V. As in modern humans, the dorsal interossei attach mainly onto the extensor expansion, the 'volar plate' and/or the respective side of the proximal phalanx of the digit to which they insert, i.e., the interosseus dorsalis 1 to the radial side of digit 2, the interosseus dorsalis 2 to the radial side of digit 3, the interosseus dorsalis 3 to the ulnar side of digit 3, and the interosseus dorsalis 4 to the ulnar side of digit 4. These four muscles thus mainly contribute to abductor these four digits.
- Usual innervation: Deep branch of the ulnar nerve (Hepburn 1892, Kallner 1956).
- Notes: The **interdigitales** present in primates such as lorisoids are usually not present in orangutans.
- Synonymy: Part of the dorsal interossei and probably part of palmar interossei *sensu* Church (1861-1862).

Flexor pollicis brevis and *flexor brevis profundus 2* (VU PP1 LSB flexor pollicis brevis + flexor brevis profundus 2, or 'superficial + deep heads of flexor pollicis brevis' = 6.1 g; VU PP2 LSB flexor pollicis brevis + flexor brevis profundus 2, or 'superficial + deep heads of flexor pollicis brevis' = 8.4 g; Figs. 33, 47, 49)

- Usual attachments: As explained above (see adductor pollicis), the flexor brevis profundus 2 of 'lower' mammals corresponds to the '**deep head of the flexor pollicis brevis**' of modern human anatomy; the '**superficial head of the flexor pollicis brevis**' of modern human anatomy thus corresponds to a true flexor pollicis brevis, being likely derived, together with the **opponens pollicis**, from the **flexor brevis profundus 1** of 'lower' mammals (for recent reviews see Diogo et al. 2009a,

Diogo & Abdala 2010 and Diogo & Wood 2012). The flexor brevis profundus 2 of orangutans usually runs mainly from the carpal region to the ulnar side of the proximal phalanx of digit 1, while the true flexor pollicis brevis runs mainly from the carpal region to the radial side of the proximal phalanx of digit 1, lying superficially to the flexor brevis profundus 2 (see Notes below).

- Usual innervation of the true flexor pollicis brevis and of the flexor brevis profundus 2: In each of the *Pongo* specimens dissected by Brooks (1887) and Hepburn (1892) the structure that they designated as 'interosseous volaris primus of Henle', which actually corresponds to the 'deep head of the flexor pollicis brevis' of human anatomy and thus to the flexor brevis profundus 2 *sensu* the present study, is innervated by the deep branch of the ulnar nerve. In each of the *Pongo* specimens dissected by Brooks (1887) and Hepburn (1892) the 'superficial head of the flexor pollicis brevis' of human anatomy (i.e., flexor pollicis brevis *sensu* the present study) is innervated by the median nerve; Kohlbrügge (1897) described a 'ulnar head' (i.e., flexor brevis profundus 2 *sensu* the present study) and a 'radial head' (i.e., flexor pollicis brevis *sensu* the present study) of the 'flexor pollicis brevis' in *Pongo*, and stated that the whole structure formed by these two 'heads' was innervated by both the median nerve and the deep branch of the ulnar nerve; Kallner (1956) reported a 'single head of the flexor pollicis brevis' (which probably corresponds to the flexor pollicis brevis plus the flexor brevis profundus 2 *sensu* the present study), and stated that this structure was innervated by the deep branch of the ulnar nerve. In our GWU PP1 specimen the true flexor pollicis brevis is innervated by the median nerve, while the flexor brevis profundus 2 is innervated by the ulnar nerve.
- Notes: As there is much confusion in the literature about the flexor brevis profundus 2 and the true flexor pollicis brevis *sensu* the present study, we will describe in some detail the configuration found in some of the orangutans dissected by us. For more details about the descriptions of other authors and for information about the synonymy, homologies and attachments of these muscles, see Diogo & Wood (2012) and see also the sections above about the adductor pollicis and the interossei palmares. In our specimens GWU PP1 and HU PP1 the true flexor pollicis originates from the flexor retinaculum and the base of metacarpal I; the flexor brevis profundus 2 originates from the base of metacarpal I and the trapezium. The true flexor pollicis brevis inserts onto the lateral side of the palmar surface of the base of the proximal phalanx of digit 1; the flexor brevis profundus 2 inserts onto the medial side of the palmar surface of the proximal phalanx of digit 1 (we could not discern if this latter structure inserts, or not, onto the sesamoid bone).
- Synonymy of the true flexor pollicis brevis: Part or totality of flexor pollicis brevis and probably part of abductor pollicis brevis (Church 1861-1862); part or totality of flexor brevis pollicis (Primrose 1899, 1900, Sonntag 1924a).

Opponens pollicis (VU PP1 LSB 4.3 g; VU PP2 LSB 4.7 g; Figs. 33, 47, 49)
- Usual attachments: From the flexor retinaculum, trapezium, adjacent sesamoid bone (e.g., Brooks 1887) and/or the tendon of the abductor pollicis brevis

(e.g., Brooks 1887, Primrose 1899, 1900) to the whole length of metacarpal I, being at least sometimes seemingly divided into a caput superficiale and a caput profundum (see Notes below).

- Usual innervation: Median nerve (Brooks 1887, Hepburn 1892).
- Notes: In Gibbs' (1999) review it is stated that in *Pongo* the opponens pollicis does not originate from the flexor retinaculum, but an origin from this structure was described by, e.g., Primrose (1899, 1900) and seemingly present in our GWU PP1 and HU PP1 specimens. Gibbs (1999) also states that the muscle is small or absent in *Pongo*, but as described by, e.g., Sullivan & Osgood (1927), in our GWU PP1 and HU PP1 specimens the opponens pollicis it actually differentiated into a small superficial slip and a well-developed deep slip that is somewhat blended with the true flexor pollicis brevis (Figs. 47, 49).

Flexor digiti minimi brevis (VU PP1 LSB 3.5 g; VU PP2 LSB 7.7 g; Figs. 33, 48–49)

- Usual attachments: From the hamate and/or the flexor retinaculum, to the ulnar side of the proximal phalanx.
- Usual innervation: Ulnar nerve (Hepburn 1892, Kallner 1956).
- Notes: Primrose (1899, 1900), Sonntag (1924a) and Kallner (1956) described an origin of the flexor digitorum minimi brevis of *Pongo* from the hamate and flexor retinaculum, and we found an origin from the flexor retinaculum in our GWU PP1 and VU PP1 specimens.
- Synonymy: A5a (Brooks 1886a); abductor digiti quinti brevis (Sullivan & Osgood 1927); flexor brevis minimi digiti (Hepburn 1892, Primrose 1899, 1900, Sonntag 1924a).

Opponens digiti minimi (VU PP1 LSB 3.3 g; VU PP2 LSB 4.3 g; Figs. 34, 48–49)

- Usual attachments: From the hamate and, often, also from the flexor retinaculum to the whole length of metacarpal V, being often divided into a superficial and a deep head (see Notes below).
- Usual innervation: Ulnar nerve (Hepburn 1892, Kallner 1956).
- Notes: As noted by Brooks (1886a), Lewis (1989) and Diogo et al. (2009a) and Diogo & Wood (2011, 2012) and corroborated by our dissections, in hominoids the opponens digiti minimi is usually slightly differentiated into superficial and deep bundles. Lewis (1989) stated that the opponens digiti minimi is more markedly divided in hominoids such as *Pan* and modern humans than in hominoids such as hylobatids, and Brooks (1886a) stated that, contrary to *Pan* and modern humans, in hominoids such as *Pongo* there are no superficial and deep bundles of the muscle separated by the deep branch of the ulnar nerve. Regarding our dissections, in hylobatids, *Gorilla* and *Pongo* the deep branch of the ulnar nerve runs mainly radial to both these bundles, and not mainly superficially (palmar) to the deep bundle and deep (dorsal) to the superficial bundle as is usually the case in *Pan* and particularly in modern humans.

- Synonymy: Opponens digiti quinti (Sullivan & Osgood 1927); opponens minimi digiti (Hepburn 1892, Sonntag 1924a).

Abductor pollicis brevis (VU PP1 LSB 6.1 g; VU PP2 LSB 9.7 g; Figs. 33, 47, 49)
- Usual attachments: From the flexor retinaculum and at least sometimes from structures such as the trapezium, the adjacent sesamoid bone, and/or the scaphoid, to the radial side of the metacarpophalangeal joint, its sesamoid bone and/or the base of the proximal phalanx of digit 1.
- Usual innervation: Median nerve (Brooks 1887, Hepburn 1892, Kallner 1956).
- Notes:
- Synonymy: Probably corresponds to part of abductor pollicis brevis *sensu* Church (1861-1862); abductor pollicis (Hepburn 1892 and Primrose 1899, 1900).

Abductor digiti minimi (VU PP1 LSB 8.2 g; VU PP2 LSB 11.0 g; Figs. 33–34, 48–49)
- Usual attachments: From the pisiform and sometimes from the flexor retinaculum and/or hamate (e.g., our specimens GWU PP1 and HU PP1), to the ulnar side of the proximal phalanx of digit 5.
- Usual innervation: Ulnar nerve (Hepburn 1892).
- Notes: Richmond (1993: Fig. 6.7) shows a 'deep head' and a 'superficial head' of the abductor digiti minimi in a *Pongo* specimen; however, these heads were not found in the orangutans reported by Primrose (1899, 1900), Hepburn (1892), Sonntag (1924a), Sullivan & Osgood (1927), Kallner (1956), and Jouffroy & Lessertisseur (1960), and dissected by us.
- Synonymy: Abductor minimi digiti (Hepburn 1892, Primrose 1899, 1900, Sonntag 1924a); abductor digiti quinti (Sullivan & Osgood 1927).

Extensor carpi radialis longus (VU PP1 LSB 36.4 g; VU PP2 LSB 44.7 g; Fig. 32, 44–46)
- Usual attachments: From the lateral supracondylar ridge and often from the lateral epicondyle of the humerus, to the base of metacarpal II and often also to base of metacarpal I (e.g., Bojsen-Møller 1978); the muscle may be fused with the extensor carpi radialis brevis (e.g., Beddard 1893).
- Usual innervation: Radial nerve (Hepburn 1892, Sonntag 1924a, Kallner 1956).
- Synonymy: Extensor carpi radialis longior (Church 1861-1861, Hepburn 1892, Beddard 1893, Primrose 1899, 1900, Sonntag 1924a).

Extensor carpi radialis brevis (VU PP1 LSB 32.0 g; VU PP2 LSB 57.0 g; Fig. 32, 44–46)
- Usual attachments: From the lateral epicondyle of the humerus and occasionally also from structures such as the radial collateral ligament (e.g., Hepburn 1892, Sonntag 1924a), to the base of metacarpal III and sometimes also to the base of metacarpal II (e.g., our specimen HU PP1).
- Usual innervation: Radial nerve (Kallner 1956).

- Notes: Michaëlis (1903) described an insertion of the extensor carpi radialis brevis onto metacarpal I in a *Pongo* specimen, but this clearly seems to be an error because Michaëlis (1903) also reported an insertion of the extensor carpi radialis longus onto metacarpal III; both insertions are completely different from the insertions found in all other orangutans dissected by us and by other authors.
- Synonymy: Extensor carpi radialis brevior (Church 1861-1862, Hepburn 1892, Beddard 1893, Primrose 1899, 1900, Sonntag 1924a).

Brachioradialis (VU PP1 LSB 263.5 g; VU PP2 LSB 254.4 g; Figs. 31–32, 40–41, 44–45)
- Usual attachments: From the shaft and distal end of the humerus to the shaft of the radius, reaching the styloid process in at least some cases (e.g., Primrose 1899, 1900).
- Usual innervation: Radial nerve (Hepburn 1892, Sonntag 1924a, Straus 1941a,b, Kallner 1956).
- Synonymy: Supinator longus (Church 1861-1862, Barnard 1875, Chapman 1878, Primrose 1899, 1900, Sonntag 1924a); supinator radii longus (Hepburn 1892, Beddard 1893).

Supinator (VU PP1 LSB 55.7 g; VU PP2 LSB 58.7 g; Figs. 31–32, 45)
- Usual attachments: From the lateral epicondyle of the humerus (caput humerale, or superficiale) and usuallly also from the proximal part of the ulna (caput ulnare, or profundum) to the proximal radius.
- Usual innervation: Posterior interosseous nerve (Hepburn 1892, Sonntag 1924a, Straus 1941a,b); deep branch of radial nerve (Kallner 1956), which passes between the two 'ulnar' heads described by her.
- Notes: In *Pongo* the origin of the supinator was from the humerus and ulna in the specimens described by Beddard (1893), Primrose (1899, 1900), Sonntag (1924a) and Sullivan & Osgood (1927) and dissected by us; it was only from the ulna in the specimens reported by Barnard (1875), Straus (1941a) and Kallner (1956).
- Synonymy: Supinator radii brevis (Barnard 1875, Beddard 1893); supinator brevis or epicondylo-radial (Hepburn 1892, Primrose 1899, 1900, Sonntag 1924a, Jouffroy 1971).

Extensor carpi ulnaris (VU PP1 LSB 26.3 g; VU PP2 LSB 44.6 g; Figs. 32, 44–46)
- Usual attachments: From the lateral epicondyle of the humerus (caput humerale) and usually also the ulna (caput ulnare) to the base of metacarpal V.
- Usual innervation: Posterior interosseous nerve (Hepburn 1892); radial nerve (Kallner 1956).
- Notes: In *Pongo* the origin of the extensor carpi ulnaris was from the humerus and ulna in the specimens described by Beddard (1893), Primrose (1899, 1900), Sonntag (1924a), Sullivan & Osgood (1927) and Straus (1941a) and dissected by us, and at least partially from the ulna in the specimens reported by Fick

(1895a,b), although Kohlbrügge (1897) and Kallner (1956) described a bony origin from the humerus only.

Anconeus (VU PP1 LSB 6.2 g; VU PP2 LSB 3.5 g; Fig. 32)

- Usual attachments: From the lateral epicondyle of the humerus to the olecranon process of the ulna and in at least some cases also to adjacent areas of the ulna (our specimens GWU PP1 and HU PP1).
- Notes: The anconeus is usually present as a distinct muscle in *Pongo* (e.g., Beddard 1893, Primrose 1899, 1900, Sonntag 1924a; our dissections).
- Synonymy: Extensor antebrachii ulnaris, epicondylo-cubitalis, anconeus brevis, anconeus parvus or anconeus quartus (Jouffroy 1971).

Extensor digitorum (VU PP1 LSB 60.0 g; VU PP2 LSB 88.2 g; Figs. 32, 44)

- Usual attachments: From the lateral epicondyle of the humerus and occasionally also from the radius (e.g., Sonntag 1924a; our specimen GWU PP1) and/or the ulna (e.g., Sonntag 1924a; our specimen GWU PP1), to the middle phalanges and (via the extensor expansions) to the distal phalanges of digits 2, 3, 4 and 5; occasionally the muscle sends more than one tendon to a digit (e.g., two tendons to digit 2 in the specimen dissected by Sonntag 1924a).
- Usual innervation: Posterior interosseous nerve (Hepburn 1892); radial nerve (Kallner 1956).
- Synonymy: Extensor communis digitorum (Church 1861-1862, Hepburn 1892, Beddard 1893, Primrose 1899, 1900, Sonntag 1924a); extensor digitorum communis (Barnard 1875, Straus 1941a,b, Kallner 1956); extensor digitorum longus (Sullivan & Osgood 1927).

Extensor digiti minimi (VU PP1 LSB 9.5 g; VU PP2 LSB 14.2 g; Figs. 32, 45–46)

- Usual attachments: From the lateral epicondyle of the humerus and/or the common extensor tendon, and/or occasionally from the ulna (e.g., Church 1861-1862, Straus 1941ab; our specimens GWU PP1 and HU PP1), to the middle phalanx and (via the extensor expansion) to the distal phalanx of digit 5; the muscle usually also inserts onto digit 4 (see Notes below).
- Usual innervation: Posterior interosseous nerve (Hepburn 1892); radial nerve (Kallner 1956).
- Notes: In the *Pongo* specimens described by Church (1861-1862), Barnard (1875), Langer (1879), Chapman (1880), Hepburn (1892), Beddard (1893), Kohlbrügge (1897), Primrose (1899, 1900), Sonntag (1924a), Sullivan & Osgood (1927), Straus (1941a), Kallner (1956), Kaneff (1980a), Aziz & Dunlap (1986) and Oishi et al. (2008, 2009) the extensor digiti minimi goes to digits 4 and 5, and according to the review of the literature done by Gibbs (1999) such an insertion onto these two digits occurs in 20 out of 23 orangutans. The **extensor digiti quarti** is usually not present as a distinct muscle in orangutans.

- Synonymy: Extensor minimi digiti (Church 1861-1862, Hepburn 1892, Beddard 1893, Primrose 1899, 1900, Sonntag 1924a); part of extensor digitorum brevis (Sullivan & Osgood 1927); extensor digiti - quarti et - quinti proprius (Straus 1941a,b); extensor digitorum lateralis, extensor digiti quinti propris, extensor digitorum secundus or extensor digiti lateralis (Kallner 1956, Kaneff 1980a, Jouffroy 1971); extensor digiti quinti-et-quarti (Aziz & Dunlap 1986); extensor digitorum proprius or profundus 4 and 5 (Lewis 1989).

Extensor indicis (VU PP1 LSB 12.2 g; VU PP2 LSB 20.6 g; Figs. 32, 45–46)

- Usual attachments: From the ulna and often also from the interosseous membrane and occasionally also from the radius (e.g., Barnard 1875) and/or the humerus/common extensor tendon (e.g., Aziz & Dunlap 1986), to digits 2 and 3; occasionally there is also an insertion onto digit 4, 5, and/or 1, and/or no insertion onto digit 3 (see Notes below).

- Usual innervation: Posterior interosseous nerve (Hepburn 1892); radial nerve (Kallner 1956).

- Notes: In *Pongo*, an insertion onto digits 2 and 3 was described by most authors, including Langer (1879), Chapman (1880), Hepburn (1892), Beddard (1893), Primrose (1899, 1900), Sullivan & Osgood (1927), Straus (1941a), Kallner (1956), Day & Napier (1963), Tuttle (1969), Tuttle & Cortright (1988), Richmond (1993) and Oishi et al. (2008, 2009) and found in our GWU PP1 specimen. In 1 of the 8 specimens dissected by Kaneff (1980a,b) and the specimen dissected by Sonntag (1924a) the muscle inserted onto digit 2 only, while Barnard (1875) described an insertion onto digits 2, 3 and 4, Church (1861-1862) described an insertion onto metacarpals II and/or III, and in our specimen HU PP1 the insertion was onto digits 1, 2 and 3. According to the review of the literature done by Straus (1941a,b), in *Pongo* an insertion on digits 2, 3, 4 and 5 occurs in about 5% of the cases, on digits 2, 3 and 4 occurs in about 11% of the cases, on digits 2 and 3 occurs in about 66% of the cases, on digit 3 occurs in about 13% of the cases, and on digit 2 occurs in about 5% of the cases. The **extensor digiti III proprius** and the **extensor brevis digitorum manus** (e.g., Diogo et al. 2009a, Diogo & Wood 2012) are usually not present as distinct muscles in *Pongo*, but the latter muscle was shown in an orangutan specimen dissected by Lewis (1989: Fig. 8.3B), being associated with the tendon of the extensor indicis to digit 3.

- Synonymy: Part or totality of extensor profundus digitorum or extensor digitorum profundus (Barnard 1875, Hepburn 1892, Wood Jones 1920, Straus 1941a,b, Kaneff 1980a); part of extensor digitorum brevis (Sullivan & Osgood 1927); extensor indicis proprius (Kallner 1956); extensor digitorum profundus proprius or extensor indicis proprius + extensor medii digiti proprius + extensor digiti quarti proprius or extensor indicis proprius + extensor medii-et-quarti (Aziz & Dunlap 1986); extensor digitorum proprius or profundus 2 and 3 (Lewis 1989).

Extensor pollicis longus (VU PP1 LSB 9.1 g; VU PP2 LSB 12.9 g; Figs. 32, 44–45)
- Usual attachments: From the ulna and interosseous membrane to the distal phalanx of the thumb and sometimes also to its proximal phalanx (e.g., Sonntag 1924a) and/or to the metacarpophalangeal join of this digit (e.g., Kallner 1956, Straus 1941a,b).
- Usual innervation: Posterior interosseous nerve (Hepburn 1892); radial nerve (Kallner 1956).
- Notes: The structure designated as '**extensor communis pollicis et indicis**' by Kaneff (1980a) is usually not present in orangutans.
- Synonymy: Extensor secundi internodii pollicis (Church 1861-1862, Hepburn 1892, Beddard 1893, Sonntag 1924a); extensor longus pollicis (Primrose 1899, 1900); extensor pollicis (Sullivan & Osgood 1927); part or totality of extensor profundus digitorum or extensor digitorum profundus (Straus 1941a,b); extensor pollicis (Aziz & Dunlap 1986); extensor digitorum proprius or profundus 1 (Lewis 1989).

Abductor pollicis longus (VU PP1 LSB 35.3 g; VU PP2 LSB 47.8 g; Figs. 32, 44–46)
- Usual attachments: From the interosseous membrane, ulna and radius, to metacarpal I and trapezium; sometimes the muscle inserts also/instead onto the scaphoid and/or sesamoid bone adjacent to the trapezium (see Notes below).
- Usual innervation:
- Notes: Apart from modern humans, in all of the primate specimens dissected by us only in hylobatids is there is a distinct **extensor pollicis brevis** that is only partially blended, proximally, with the belly of the abductor pollicis longus (e.g., Diogo & Wood 2011, 2012). It should be noted that some authors have described an 'extensor pollicis brevis' and an 'abductor pollicis longus' in primate taxa other than hylobatids and modern humans, including chimpanzees (see Synonymy below). For example, in gorillas the name 'extensor pollicis brevis' has been used (e.g., Hepburn 1892, Straus 1941a,b) to refer to a tendon of the abductor pollicis longus (*sensu* the present study) that inserts onto the proximal phalanx of the thumb (i.e., to the typical insertion point of the extensor pollicis brevis of modern humans). However, as stressed by authors such as Kaneff (1979, 1980a,b) and Aziz & Dunlap (1986) and corroborated by our dissections in gorillas there is usually a single fleshy belly of the abductor pollicis longus that then gives rise to the so-called 'tendons of the extensor pollicis brevis and of the abductor pollicis longus'; this configuration is usually also the case in *Pongo* and *Pan*. Thus, contrary to the condition in *Homo* and hylobatids, in *Pongo*, *Pan* and *Gorilla* the extensor pollicis brevis is usually *not* present as a separate muscle. In *Pongo* the insertion of the abdutor pollicis longus is to the sesamoid and metacarpal I according to Hepburn (1892) and Sonntag (1924a), to the sesamoid, trapezium and metacarpal I according to Primrose (1899, 1900), to the trapezium and metacarpal I according to Church

(1861-1862), Hartmann (1886), Beddard (1893), Kohlbrügge (1897), Straus (1941a,b), Aziz & Dunlap (1986) and to our dissections, to the trapezium and the scaphoid according to Fick (1895a,b), and to the metacarpal I according to Kallner (1956). The insertion onto the trapezium and metacarpal I represents the common condition for *Pongo* according to Gibbs (1999), and according to the review of the literature done by Straus (1941a,b) an insertion onto the proximal phalanx of the thumb occurs in no orangutans. In agreement to what is shown in Lewis' (1989) Fig. 8.3B, the distal part of the abductor pollicis longus of our specimens GWU PP1 and HU PP1 gives rise to two tendons, which probably correspond to the distal tendons of the abductor pollicis longus and of the extensor pollicis brevis of humans and hylobatids; however none of these tendons inserts onto the proximal phalanx of digit 1.

- Synonymy: Extensor ossis metacarpi pollicis (Hepburn 1892, Primrose 1899, 1900, Wood Jones 1920); extensor ossis metacarpi pollicis plus extensor primi internodii pollicis (Beddard 1893, Sonntag 1924a); extensor ossis metacarpi pollicis plus extensor pollicis brevis (Primrose 1899, 1900); abductor pollicis longus plus extensor pollicis brevis (Kallner 1956); supinator manus or extensor metacarpi pollicis (Howell 1936a,b, Jouffroy 1971).

CHAPTER **5**

Trunk and Back Musculature

Obliquus capitis inferior (VU PP1 LSB = 5.6 g; Fig. 17)
- Usual attachments: From the spinous processes of C2 to the transverse process of C1 (Sonntag 1924a).
- Usual innervation: Data are not available.

Obliquus capitis superior (VU PP1 LSB = 4.5 g; Fig. 17)
- Usual attachments: From the transverse process of C1 to the occipital bone (Fig. 17).
- Usual innervation: Data are not available.

Rectus capitis anterior
- Usual attachments: Rectus capitis anterior major runs from C7 and T1 to T5 in common with the longus colli, to the basiocciput; the rectus capitis anterior minor runs from the anteroventral surface of the lateral mass of C1 to the basiocciput, ventral to the foramen magnum and occipital condyle, and dorsolateral to the insertion of the longus capitis (Sonntag 1924a).
- Usual innervation: Data are not available.

Rectus capitis lateralis
- Usual attachments: From the transverse process of C1 to the jugular process of the occipital bone, dorsal to the jugular foramen and lateral to the occipital condyle (Dean 1984); the muscle may however be missing in *Pongo* (Sonntag 1924a).
- Usual innervation: Data are not available.

Rectus capitis posterior major (VU PP1 LSB = 14.6 g; Fig. 16)
- Usual attachments: From the spinous process of C2 to the occipital bone, between the inferior nuchal line and the foramen magnum, lateral to the insertion of the rectus capitis posterior minor (Sonntag 1924a).
- Usual innervation: Data are not available.

Rectus capitis posterior minor (VU PP1 LSB = 4.8 g; Fig. 17)
- Usual attachments: From the dorsal tubercle process of C1 to the medial region of the inferior nuchal line (Sonntag 1924a).
- Usual innervation: Data are not available.

Longus capitis (VU PP1 LSB = 15.7 g; Fig. 11)
- Usual attachments: From cervical vertebrae, to the basiocciput (Dean 1985).
- Usual innervation: Data are not available.

Longus colli (VU PP1 LSB = 11.2 g; Fig. 11)
- Usual attachments: The superior oblique division is usually missing in *Pongo*, while the inferior oblique division and the vertical division originate from the body of C6-T5; the muscle inserts onto C1 and C6 (Sonntag 1924a).
- Usual innervation: Data are not available.

Scalenus anterior (Fig. 9)
- Usual attachments: From the transverse processes of C3–C6 to the anterior scalene tubercle of the first rib (Sonntag 1924a).
- Usual innervation: Data are not available.
- Notes: Stewart (1936) stated that orangutans and other apes usually have a scalenus anterior and a 'scalenus medius' (which corresponds to the scalenus medius + scalenus posterior *sensu* the present study). According to him the scalenius anterior usually inserts onto the first rib in primates, its origin being limited to C5 and C6 in 'lower primates', but in hominoids the insertion included some of the upper cervical vertebrae. Also according to him, within primates the 'scalenus medius' (i.e., medius + posterior *sensu* the present study) shows a progressive decrease in the number of ribs supplying attachment, usually inserting in hominoids upon the first rib, but occasionally also onto the second rib.

Scalenus medius (Fig. 9)
- Usual attachments: Mainly from C3–C4 to first rib, the muscle being partly fused with the scalenus posterior (Sonntag 1924a).
- Usual innervation: Data are not available.
- Notes: See notes about scalenus anterior, above.

Scalenus posterior (Fig. 9)
- Usual attachments: The combined scalenus medius-scalenus posterior originates from C2–C6; the scalenus posterior inserts mainly onto the first rib (Sonntag 1924a).
- Usual innervation: Data are not available.
- Notes: See notes about scalenus anterior above. The **scalenus minimus** is usually not present as a distinct muscle in orangutans.

Levatores costarum
- Usual attachments: Data are not available.
- Usual innervation: Data are not available.

Intercostales externi (Fig. 14)
- Usual attachments: Attached to the adjacent margins of each pair of ribs (Fig. 14).
- Usual innervation: Data are not available.

Intercostales interni
- Usual attachments: Connect the adjacent margins of each pair of ribs (our specimens).
- Usual innervation: Data are not available.

Transversus thoracis
- Usual attachments: Data are not available.
- Usual innervation: Data are not available.

Splenius capitis (VU PP1 LSB = 69.0 g; Fig. 5, 13, 15–16, 35–37)
- Usual attachments: From the spinous processes of C2–T4 and the supraspinous ligaments, to the mastoid process and the occipital bone beneath the superior occipital line (Sonntag 1924a).
- Usual innervation: Data are not available.

Splenius cervicis (VU PP1 LSB = 32.0 g)
- Usual attachments: The muscle was missing in the *Pongo* specimen reported by Sonntag (1924a), but was present in the only *Pongo* specimen where we analysed this region of the body in detail, i.e., VU PP1, running mainly from the thoracic to the cervical vertebrae.
- Usual innervation: Data are not available.

Serratus posterior superior
- Usual attachments: According to Sonntag (1924a) the muscle is absent or represented by tendinous threads in *Pongo*.
- Usual innervation: Data are not available.

Serratus posterior inferior (Fig. 14)
- Usual attachments: From the thoracolumbar fascia to the inferior four or five ribs, lateral to their angles (Sonntag 1924a).
- Usual innervation: Data are not available.

Iliocostalis (Fig. 14)
- Usual attachments: The costal attachments are onto all the ribs, the muscle being fused with the longissimus and also originating from the transverse processes of the lumbar vertebrae and from T1–T4; insertion is onto C4–C6 and onto ribs 1–12 (Sonntag 1924a).
- Usual innervation: Data are not available.
- Notes: The iliocostalis, longissimus and spinalis form the '**erector spinae**' (see Fig. 14).

Longissimus (Fig. 14)
- Usual attachments: The longissimus is fused with the iliocostalis and semispinalis capitis; the cranial insertion of the longissimus is to the occiput, the cervical insertion is onto the transverse processes of C2–C6, and the thoracic insertion is onto the ribs between the costal angle and the transverse processes of the thoracic vertebrae.
- Usual innervation: Data are not available.
- Notes: The iliocostalis, longissimus and spinalis form the '**erector spinae**' (see Fig. 14). In chimpanzees (e.g., Gratiolet & Alix 1866), modern humans, gorillas and a few other primates, but seemingly not in orangutans, the **atlantomastoideus** might be occasionally present as a distinct muscle, running from the atlas to the mastoid process (Diogo & Wood 2012).

Spinalis (Fig. 14)
- Usual attachments: Mainly connects the spinous processes of different vertebrae (Fig. 14).
- Usual innervation: Data are not available.
- Notes: The iliocostalis, longissimus and spinalis form the '**erector spinae**' (see Fig. 14).

Semispinalis thoracis (Fig. 18)
- Usual attachments: The origin of the muscle is separable from the origin of semispinalis cervicis, from the transverse processes of T6–T12; insertion is onto the spinous processes of C6–T4 (Sonntag 1924a).
- Usual innervation: Data are not available.

Semispinalis cervicis (Figs. 16–18)
- Usual attachments: From the transverse processes of T1–T4 and the articular processes of C4–C7, to the spinous processes of C2–C5 (Sonntag 1924a).
- Usual innervation: Data are not available.

Semispinalis capitis (VU PP1 LSB = 55.8 g; Fig. 16)
- Usual attachments: From the articular processes of C3–T6 and the spinous process of C7, to the occipital bone between the superior and inferior nuchal lines (Sonntag 1924a). In the *Pongo* specimen reported by Sonntag (1924a), the muscle also has, on its internal surface, a separate slender Y-shaped bundle of fibres, the vertical limb arising from fascia covering the muscle and the two diverging limbs blending with the main body of the muscle before its insertion onto the occipital bone (Sonntag 1924a). According to Gibbs (1999), this bundle of fibers corresponds to the '**biventer cervicis**', which is usually blended with the semispinalis capitis or be present as a distinct muscle in modern humans.
- Usual innervation: Data are not available.

Multifidus (Fig. 18)
- Usual attachments: This muscle is shown in Fig. 18; to our knowledge it has never been shown or described in detail in *Pongo*.
- Usual innervation: Data are not available.

Rotatores (Fig. 18)
- Usual attachments: These muscles are shown in Fig. 18; they mainly connect the transverse processes of the thoracic vertebrae to the spinous process of the vertebra above; to our knowledge it has never been shown or described in detail in *Pongo*.
- Usual innervation: Data are not available.
- Synonymy: Rotatores breves and longi (Gibbs 1999).

Interspinales
- Usual attachments: To our knowledge, there are no detailed descriptions of these muscles in orangutans.
- Usual innervation: Data are not available.

Intertransversarii
- Usual attachments: To our knowledge, there are no detailed descriptions of these muscles in orangutans, the only mention to them being Sonntag (1924a), who stated that they are similar to those of modern humans.
- Usual innervation: Data are not available.

CHAPTER 6

Diaphragmatic and Abdominal Musculature

Diaphragma (VU PP1 680.5 g unpaired muscle, so each side = 340.25 g; Fig. 24)
- Usual attachments: This muscle is shown in Fig. 24; to our knowledge, it has never been shown or described previously, in *Pongo*.
- Usual innervation: Data are not available.

Rectus abdominis (VU PP1 LSB = 284.9 g; Fig. 20)
- Usual attachments: To our knowledge, it has never been shown or described previously, in *Pongo*. According to our dissections it mainly runs from the outer surface of the costal cartilages to the region of the pubis crest, and has four tendinous intersections (Fig. 20).
- Usual innervation: Data are not available.
- Notes: To our knowledge, the presence of a **supracostalis** and a **tensor linea semilunaris** has not been reported in orangutans.

Pyramidalis
- Usual attachments: To our knowledge, it has never been shown or described previously, in *Pongo*.
- Usual innervation: Data are not available.

Cremaster
- Usual attachments: From the internal abdominal oblique to the 'inguinal ligament' (see Notes below), also receiving a contribution from the transversus abdominis (Mijsberg 1923, Miller 1947, Lunn 1949).
- Usual innervation: Data are not available.
- Notes: According to the literature review of Gibbs (1999) there is no true inguinal ligament in any ape, only a series of tendinous arches over the sartorius and the femoral vessels and nerves, merging with the fascia lata (e.g., Miller 1947, Lunn 1948). Miller (1947) and Lunn (1948) suggest that a true inguinal ligament is only found in modern humans, but ligament-like connective tissue in a plane ventral to the pelvic girdle has been identified in all of the mammals dissected by Lunn (1948).

Obliquus externus abdominis (VU PP1 LSB = 291.4 g; Figs. 19, 21, 25–26, 29)
- Usual attachments: Originates by fleshy slips from the external surface of ribs and inserts onto the pelvic region (Figs. 19, 21, 25-26, 29).
- Usual innervation: Data are not available.

Obliquus internus abdominis (VU PP1 LSB = 222.2 g; Fig. 22)
- Usual attachments: It forms the conjoint tendon with the transversus abdominis, which is described below.
- Usual innervation: Data are not available.

Transversus abdominis (VU PP1 LSB = 129.3 g; Fig. 23)
- Usual attachments: This muscle forms the posterior layer of the rectus sheath (e.g., Lunn 1949). Its fibres decussate in the linea alba, and the aponeurotic part forms, along with the internal oblique, the conjoined tendon, which inserts into the superior pubic surface in the region of the pubic crest (Miller 1947, Lunn 1949).
- Usual innervation: Data are not available.

Quadratus lumborum (VU PP1 LSB = 703.0 g; Fig. 24)
- Usual attachments: To our knowledge this muscle has never been described in detail or shown in *Pongo*; it mainly from the iliac region to the lumbar vertebrae (Fig. 24).
- Usual innervation: Data are not available.

CHAPTER 7

Perineal, Coccygeal and Anal Musculature

Coccygeus
- Usual attachments: The muscle is mostly replaced by the strong sacrospinous ligament.
- Usual innervation: Data are not available.

Iliococcygeus
- Usual attachments: Origin is usually aponeurotic, from the obturator fascia, extending as far as the ischial spine; the muscle is continuous with the sacrosciatic ligaments and the piriformis, and inserts onto the coccyx (Gibbs 1999).
- Usual innervation: Data are not available.

Levator ani
- Usual attachments: The muscle is formed by bilateral 'plates' of muscle and fibrous tissue that have a linear attachment to the obturator fascia on the inner wall of the lesser pelvis that extends from the pubic symphysis to the ischial ramus; the posteriorly directed fibres converge in the midline; the more anterior fibres encircle the rectum and the more posterior ones insert onto the anococcygeal raphe behind the anus as well as onto the tip of the coccyx (Gibbs 1999).
- Usual innervation: Data are not available.
- Notes: To our knowledge, the ***pubovesicalis*** ('ligamentum puboprostaticum' or 'puboampullaris': Gibbs 1999) has never been reported in orangutans. Remnants of the **flexor caudae** have however been reported in *Pongo* (Lartschneider 1895, Eggeling 1896).

Pubococcygeus
- Usual attachments: Origin is from the posterior surface of the pubic body, extending to the ischial spine and also to the obturator fascia; insertion is onto the rectal wall and the tip of the coccyx; the pubococcygeus muscles of the two sides unite at the midline (Thompson 1901, Smith 1923, Elftman 1932).

- Usual innervation: Data are not available.
- Notes: According to Elftman (1932) and Thompson (1901) in great apes the *puborectalis* is homologous with the inferior fibres of the pubococcygeus.

Sphincter ani externus

- Usual attachments: The sphincter encircles the anus, running to the perineal body and forming a muscular basin (less marked degree in *Pan* than in *Gorilla* and *Pongo*) for the support of both alimentary and urogenital viscera; some fibres radiate into the raphe of the bulbospongious (Elftman 1932).
- Usual innervation: Data are not available.

Bulbospongiosus

- Usual attachments: Origin is from the median raphe of the penile bulb and the perineal body, the muscle surrounding the bulb and the corpora of the penis, and its superior fibres merging with those of the external anal sphincter; the bulbospongious inserts onto the inferior surface of the penis (Elftman 1932).
- Usual innervation: Data are not available.
- Notes: To our knowledge, there are no descriptions of the **transversus perinei profundus, transversus perinei superficialis, ischiocavernosus, rectococcygeus** (caudoanalis), **rectourethralis, rectovesicalis, regionis analis, regionis urogenitalis, compressor urethrae, sphincter urethrovaginalis, sphincter pyloricus, suspensori duodeni, sphincter ani internus, sphincter ductus choledochi, sphincter ampullae, detrusor vesicae, trigoni vesicae, vesicoprostaticus, vesicovaginalis, puboprostaticus, rectouterinus** and the **sphincter urethrae** in *Pongo*. These muscles clearly need to be studied in detail in orangutans.
- Synonymy: Bulbocavernosus (Gibbs 1999).

CHAPTER 8

Pelvic and Lower Limb Musculature

Iliacus (Figs. 24, 52-53)
- Usual attachments: From the iliac fossa (Hepburn 1892) and/or the entire ventral surface of the ilium (Sigmon 1974) to the medial aspect of the lesser trochanter of the femur, in combination with the psoas major (Beddard 1893, Sigmon 1974).
- Usual innervation: Branches of femoral nerve (Sigmon 1974).

Psoas major (VU PP1 LSB = 110.1 g; Figs. 24, 51–53)
- Usual attachments: From the lateral surfaces of the bodies and the costal processes of the lumbar vertebrae, extending proximally to T12 and distally to S1; it also takes origin from the intervening intervertebral discs, and joins with the iliacus to insert as iliopsoas onto the lesser trochanter and distally on the adjacent shaft (Champneys 1872, Hepburn 1892, Boyer 1935b, Sigmon 1974).
- Usual innervation: Femoral nerve and first two or three lumbar nerves (Hepburn 1892, Sonntag 1924a, Sigmon 1974).

Psoas minor (VU PP1 LSB = 34.0 g; Fig. 24)
- Usual attachments: Originates mainly from the ventrolateral surface of L1 and the last thoracic vertebra (Hepburn 1892, Sigmon 1974). An origin from L2 (Hepburn 1892) and/or from the intervertebral discs (Sigmon 1974) may also be present. Insertion is onto the iliopubic eminence and the pectineal line (Hepburn 1892, Sigmon 1974).
- Usual innervation: First lumbar nerve and often the twelfth thoracic nerve (Hepburn 1892, Sigmon 1974).

Gluteus maximus (VU PP1 LSB = 232.2 g, RSB = 241.36; Figs. 12, 50–51)
- Usual attachments: This is a thin, flat muscle, with its proximal portion thinner than the distal one (Hepburn 1892, Owen 1830-1831, Sigmon 1974, Sonntag 1924a, Zihlman & Brunker 1979, Ferrero 2011, Ferrero et al. 2012). It is located on the superficial layer of the gluteal region, covering the distal portion of the gluteus medius, its fibers running obliquely and being coated by much fat. Origin

is from the iliac crest, thoracolumbar fascia, sacrum, coccyx, the sacrotuberal ligament and the fascia over gluteus medius (Beddard 1893, Boyer 1935b, Van den Broek 1914, Brown 1983, Champneys 1872, Ferrero 2011, Ferrero et al. 2012, Hamada 1985, Hepburn 1892, Miller 1945, Mysberg 1917, Ranke 1897, Robinson et al. 1972, Sigmon 1974, Sonntag 1924a). There is also an origin from the ischial tuberosity (Beddard 1893, Boyer 1935b, Brown 1983, Hamada 1985, Hepburn 1892, Robinson et al. 1972, Sigmon 1974, Sonntag 1924a, Zihlman & Brunker 1979) and often from the sacrospinous ligament (Ferrero 2011, Ferrero et al. 2012, Sigmon 1974, Sonntag 1924a). Insertion is onto the iliotibial tract (when present) and onto the posterolateral aspect of the femur in the region of the gluteal tuberosity (Appleton & Ghey 1929, Boyer 1935b, Brown 1983, Chapman 1880, Ferrero 2011, Ferrero et al. 2012, Hamada 1985, Sigmon 1974, Sonntag, 1924a). It may also insert onto the hypertrochanteric fossa, on the lateral aspect of the femur (Appleton & Ghey 1929). The insertion of the muscle is generally more distal on the femur in apes than in *Homo*, extending almost two-thirds down the femur in *Pongo* (Hamada 1985, Sigmon 1974, Stern 1972) and one-third down the femur in our specimen GWU PP1. The gluteus maximus may be divided into two bellies (Brown 1983; Hepburn 1892, Sonntag 1924a) and/or fused with the distal portion of the biceps femoris (Sonntag 1924a); in our specimen VU PP1 it is fused with the ischiofemoralis. The ischiofemoral head of the biceps femoris described in *Pongo* corresponds to a distal head of the gluteus maximus (Preuschoft 1961, Sonntag 1924a, Stern 1972).

- Usual innervation: Inferior gluteal nerve (Sigmon 1974).

Gluteus medius (VU PP1 LSB = 225.95 g, RSB = 206.41 g; Fig. 50)

- Usual attachments: This is a triangular shape muscle, thinner on its cranial portion and thicker on the caudal one. It is located superficially to the gluteus minimus and lateral to the piriformis. Origin is from the lateral surface of the ilium and the gluteal fascia (Beddard 1893, Boyer 1935b, Sigmon 1974) and occasionally from the fascia lata (Beddard 1893). Insertion is onto the lateral aspect of the greater trochanter (Boyer 1935b). Our specimen VU PP1 has an insertion extending to the lateral portion of the vastus lateralis. Some authors described a fusion with the piriformis (Boyer 1935b) but in our specimen VU PP1 the two muscles were clearly separate, the gluteus medius being instead fused with the tensor fasciae latae.
- Usual innervation: Superior gluteal nerve (Sigmon 1974).

Gluteus minimus (VU PP1 LSB = 20.89 g, RSB = 17.56 g; Fig. 50)

- Usual attachments: This is a fan-shaped thin muscle (Sigmon 1974), its fibers running obliquely to the sagittal axis of the body. Origin is from the ilium, extending from just distal to the anterior superior iliac spine, towards the acetabulum (Boyer 1935b, Chapman 1880, Sigmon 1974, Sonntag 1924a), from the margin of the greater sciatic notch (Beddard 1893, Sigmon 1974). The origin extends to the ischial spine (Boyer 1935b, Sigmon 1974) and may also extend to the sacrospinous ligament (Boyer 1935b). Insertion is onto the greater trochanter (Hepburn 1892, Sigmon 1974, Sonntag 1924a), occasionally extending onto the capsule of the femoral head (Beddard 1893).

- Usual innervation: Superior gluteal nerve (Sigmon 1974).

Scansorius (VU PP1 LSB = 18.6 g, RSB = 40.53 g; Fig. 50)

- Usual attachments: This is a flat triangular shape muscle (Beddard 1893, Boyer 1935b) located lateral to the ilium, deeper to the gluteus medius and gluteus minimus. Origin is from the ilium adjacent to the acetabular rim (Beddard 1893, Boyer 1935b), extending from the anterior superior iliac spine to the anterior inferior iliac spine (Boyer 1935b, Pearson & Davin 1921). Insertion is onto the greater trochanter, distal to that of gluteus minimus (Beddard 1893, Boyer 1935b, Hepburn 1892, Owen 1830-1831, Sigmon 1974). The scansorius may be fused with the iliacus (Sigmon 1974).
- Usual innervation: Superior gluteal nerve (Boyer 1935b, Sigmon 1974).

Ischiofemoralis (VU PP1 LSB = 148.4 g, RSB = 162.24 g; Figs. 50, 51, 53)

- Usual attachments: Although initially described by some authors as comprising the distal third of the gluteus maximus (Stern 1972, Sigmon 1974, Hamada 1985), the ischiofemoralis is now considered to be an independent muscle. This muscle is usually missing in modern humans but it is particularly developed in *Pongo*. Origin is from the ischial tuberosity, the muscle sharing a common tendinous aponeurosis with the biceps femoris and the semitendinosus (Brown 1983). The most proximal fibres are fused to the gluteus maximus and insert directly onto the proximal femoral shaft; the most distal fibres penetrate the tensor fascia latae and insert onto the aponeurosis of the vastus lateralis (Brown 1983, Ferrero 2011, Ferrero et al. 2012).
- Usual innervation: Inferior gluteal nerve (Brown 1983).

Gemellus superior (VU PP1 LSB = 2.78 g, RSB = 2.79 g; Fig. 51)

- Usual attachments: Sigmon (1974) stated that the muscle is absent in some *Pongo*, but in the two specimens where we analysed this region of the body in detail the muscle was present (GWU PP1 and VU PP1). Origin is from the region of the ischial spine between the ischial spine and the ischial tuberosity (Boyer 1935b, Sigmon 1974). It has a common insertion onto the trochanteric fossa with the tendon of obturator internus, with which it is fused (Beddard 1893, Boyer 1935b, Ferrero 2011, Ferrero et al. 2012, Sigmon 1974, Sonntag 1924a).
- Usual innervation: Branches of sacral plexus (Sigmon 1974).

Gemellus inferior (VU PP1 LSB = 4.94 g, RSB = 4.07 g; Fig. 51)

- Usual attachments: Origin is from the region of the ischial tuberosity (Boyer 1935b, Hepburn 1892); close to its insertion it fuses with the inferior border of

obturator internus (Boyer 1935b, Hepburn 1892, Sigmon 1974), and then the two muscles insert onto the trochanteric fossa (Beddard 1893, Sigmon 1974, Sonntag 1924a). Hepburn (1892) reported that the muscle is blended with the quadratus femoris, but in our specimens GWU PP1 and VU PP1 it is instead blended with the obturator externus and the gemellus superior.
- Usual innervation: Branches of sacral plexus (Sigmon 1974).

Obturatorius externus (VU PP1 LSB = 22.27 g, RSB = 21.74 g)
- Usual attachments: Runs from the external surface of the medial bony margin of the obturator foramen and from the obturator membrane (Boyer 1935b, Sigmon 1974), to the trochanteric fossa (Boyer 1935b, Ferrero 2011, Ferrero et al. 2012, Hepburn 1892, Sigmon 1974, Sonntag 1924a). It is usually fused with the obturator internus (Beddard 1893, Hepburn 1892) and in our specimen VU PP1 it is fused with both gemelli.
- Usual innervation: Obturator nerve (Sigmon 1974).
- Synonymy: Obturator externus (Gibbs 1999).

Obturatorius internus (VU PP1 LSB = 39.24 g, RSB = 38.41 g; Fig. 51)
- Usual attachments: From the margin of the obturator foramen and the obturator membrane as well as from the superior portion of the descending pubic ramus and the medial surface of the inferior ischial ramus (Boyer 1935b, Sigmon 1974), to the trochanteric fossa (Beddard 1893, Boyer 1935b, Sigmon 1974).
- Usual innervation: Sacral nerves (Sigmon 1974).
- Synonymy: Obturator internus (Gibbs 1999).

Piriformis (VU PP1 LSB = 26.49 g, RSB = 26.54 g; Fig. 50)
- Usual attachments: It originates by slips from the distal half of the sacrum (Boyer 1935b, Hepburn 1892, Sigmon 1974). The distant extent of the origin is S5 (Hepburn 1892), and an origin from the greater sciatic notch may be present (Chapman 1880, Sigmon 1974). In a single specimen of *Pongo* the sacral portion of the origin was absent (Chapman 1880). The piriformis descends thought the greater sciatic foramen to an insertion onto the anteromedial aspect of the tip of the greater femoral trochanter (Beddard 1893, Boyer 1935b, Chapman 1880, Hepburn 1892, Sigmon 1974). The muscle is usually fused with the gluteus medius (Beddard 1893, Boyer 1935b, Sigmon 1974, Sonntag 1924a).
- Usual innervation: Branches of sacral plexus (Sigmon 1974).

Quadratus femoris (VU PP1 LSB = 33.55 g, RSB = 33.63 g; Fig. 51)
- Usual attachments: From the anterolateral aspect of the ischial tuberosity (Beddard 1893, Boyer 1935b, Miller 1945, Sigmon 1974, Sonntag 1924a), although it may extend medially to this tuberosity (Boyer 1935b), to the intertrochanteric crest (Boyer 1935b, Hepburn 1892, Sigmon 1974) and often the greater trochanter (Hepburn 1892, Sonntag 1924a) and/or the lesser

trochanter (Hepburn 1892, Sonntag 1924a), fusing with pectineus in one specimen of *Pongo* (Sonntag 1924a).
- Usual innervation: Sacral nerves (Sigmon 1974).
- Notes: Hepburn (1892) stated that the **articularis genu** is present in all apes, but the muscle is usually not reported in orangutans and in the specimens in which we examined this anatomical region in detail (GWU PP1 and VU PP1) the articularis genu is seemingly not present as a distinct muscle.

Rectus femoris (VU PP1 LSB = 95.57 g, RSB = 93.72 g; Figs. 50–53)
- Usual attachments: The rectus femoris has two heads in 1 out of 5 orangutans according to Gibbs (1999); in the orangutans where we dissected this anatomical region in detail there is only a short head, the reflected head being missing (GWU PP1 and VU PP1). In one *Pongo* specimen there was a single head but this head was divided by a fibrous septum (Sonntag 1924a). The short head, usually present, is originated from the anterior inferior iliac spine (Boyer 1935b, Ferrero 2011, Ferrero et al. 2012, Sigmon 1974, Sonntag 1924a). The reflected head, when present, is originated from the ilium, superior to the acetabulum (Beddard 1893, Boyer 1935b, Hepburn 1892, Sigmon 1974). Insertion of the rectus femoris is onto the patella together with the vasti muscles, and then via the patellar tendon to the tibial tuberosity (Beddard 1893, Boyer 1935b, Hepburn 1892, Sonntag 1924a).
- Usual innervation: Data are not available.

Vastus intermedius (VU PP1 LSB = 29.29 g, RSB = 47.63 g)
- Usual attachments: From the ventral femoral shaft (Ferrero 2011, Ferrero et al. 2012, Sigmon 1974) to the patella together with the rectus femoris and the other vasti muscle, and then via the patellar tendon to the tibial tuberosity (Beddard 1893, Boyer 1935b, Hepburn 1892, Sonntag 1924a).
- Usual innervation: Data are not available.

Vastus lateralis (VU PP1 LSB = 133.88 g, RSB = 131.42 g; Figs. 51, 53)
- Usual attachments: From the lateral aspect of the greater femoral trochanter and the distal two-thirds of the lateral femoral shaft, near the lateral lip of the linea aspera (Beddard 1893, Boyer 1935b, Sigmon 1974, Sonntag 1924a), and occasionally also from the iliofemoral ligament (Sigmon 1974), to the patella together with the rectus femoris and the other vasti muscles, and then via the patellar tendon to the tibial tuberosity (Beddard 1893, Boyer 1935b, Hepburn 1892, Sonntag 1924a).
- Usual innervation: Data are not available.

Vastus medialis (VU PP1 LSB = 66.39 g, RSB = 65.61 g; Figs. 51–53)
- Usual attachments: From the posteromedial femoral shaft, near the region of the linea aspera (Beddard 1893, Boyer 1935b, Ferrero 2011, Ferrero et al. 2012, Sigmon 1974, Sonntag 1924a) and often also from the iliofemoral ligament (Beddard 1893, Ferrero 2011, Ferrero et al. 2012, Sigmon 1974), attaching

onto the intermuscular septa (Sonntag 1924a) and inserting onto the patella together with the rectus femoris and the other vasti muscle, and then via the patellar tendon to the tibial tuberosity (Beddard 1893, Boyer 1935b, Hepburn 1892, Sonntag 1924a).

- Usual innervation: Data are not available.

Sartorius (VU PP1 LSB = 38.91 g, RSB = 38.61 g; Figs. 51–53, 59)

- Usual attachments: From the ventral ('anterior') iliac border (Beddard 1893, Hepburn 1892), often including the anterior superior iliac spine (Boyer 1935b, Ferrero 2011, Primrose 1899, Sigmon 1974, Sonntag 1924a), and occasionally also from the lateral iliac border (Sigmon 1974; our specimen VU PP1), to the medial border of the tibial shaft (Beddard 1893, Boyer 1935b, Ferrero 2011, Ferrero et al. 2012, Hepburn 1892, Sigmon 1974, Sonntag 1924a). The insertion of the sartorius is superficial to those of gracilis and semitendinosus (Beddard 1893, Boyer 1935b, Ferrero 2011, Ferrero et al. 2012, Hepburn 1892).
- Usual innervation: Femoral nerve (Sigmon 1974).

Tensor fasciae latae (VU PP1 LSB = 30.05 g, RSB = 14.56 g)

- Usual attachments: This muscle is present in all hominoids but it is often said that in *Pongo* the muscle is usually missing or rather small (Chapman 1880, Kaplan 1958a); it has been suggested that it may have been replaced functionally by the scansorius (Sigmon 1974). The tensor fasciae latae is however present in one of the specimens where we analyses this anatomical region in detail (VU PP1), being absent in GWU PP1. When present the muscle runs from the anterior superior iliac spine to the iliotibial tract (Chapman 1880, Duvernoy 1855-1856, Ferrero 2011, Ferrero et al. 2012, Hepburn 1892, Kaplan 1958a, Sigmon 1974), being often fused proximally with the gluteus maximus (Hepburn 1892, Sigmon 1974) and laterally with the gluteus medius and the gluteus minimus (Sigmon 1974).
- Usual innervation: When present, it is innervated by the superior gluteal nerve (Sigmon 1974).

Adductor brevis (VU PP1 LSB = 35.84 g, RSB = 43.72 g; Fig. 53)

- Usual attachments: Origin is from the body of the pubis (Beddard 1893, Hepburn 1892, Sigmon 1974), often also from the inferior pubic ramus near the symphysis (Beddard 1893, Sonntag 1924a), and in our specimen VU PP1 it was also from the distal region of the pubic tubercle, and occasionally may be also from the superior pubic ramus (Boyer 1935b, Sigmon 1974; our specimens GWU PP1 and VU PP1). In our specimens GWU PP1 and VU PP1 the adductor brevis was divided into two portions, one taking origin from the superior pubic ramus and the other from the inferior pubic ramus (Ferrero 2011, Ferrero et al. 2012); such a configuration was reported as a variation in *Pongo* by other authors (Boyer 1935b, Hepburn 1892, Sonntag 1924a). The insertion of the adductor brevis is distal to the lesser trochanter and onto the superior third of the medial lip of the linea aspera, onto the mid-dorsal femoral surface, onto the pectineal line (Appleton & Ghey 1929, Beddard 1893, Boyer 1935b, Hepburn 1892, Sigmon 1974). The adductor brevis is partly fused with the short head of the adductor magnus in some orangutans (Hepburn 1892, Sigmon 1974).

- Usual innervation: Ventral ('anterior') division of the obturator nerve (Sigmon 1974, Hamada 1985).

Adductor longus (VU PP1 LSB = 60.94 g, RSB = 61.64 g; Figs. 52, 59)

- Usual attachments: It originates by a flat tendon from the anterior superior pubic ramus in the region of the pubic tubercle (Beddard 1893, Sigmon 1974, Sonntag 1924a) and occasionally from the superior margin of the pectineus (Boyer 1935b). Insertion is onto the middle of the medial lip of the linea aspera or mediodorsal femoral shaft (Appleton & Ghey 1929, Beddard 1893, Boyer 1935b, Ferrero 2011, Ferrero et al. 2012, Hepburn 1892, Sigmon 1974, Sonntag 1924a), besides and ventral to the proximal half of the insertion of the short head of adductor magnus (Beddard 1893, Sigmon 1974). The adductor longus may be fused with the vastus medialis (Beddard 1893; Sonntag 1924a).
- Usual innervation: Data are not available.

Adductor magnus (VU PP1 LSB = 400.53 g, RSB = 369.5 g; Figs. 53, 59)

- Usual attachments: Origin is from the ventral surface of the inferior pubic ramus, lateral to the symphysis, and from the inferior ischial ramus as far as the ischial tuberosity (Beddard 1893, Boyer 1935b, Hepburn 1892, Sigmon 1974, Sonntag 1924a, Yirga 1987). This origin is described as continuous (Beddard 1893, Hepburn 1892, Sigmon 1974, Sonntag 1924a), although this was not the case in our specimen VU PP1, in which there were two heads of origin, one from the inferior pubic ramus and the other from the ischial ramus (Ferrero 2011, Ferrero et at. 2012). Occasionally the adductor magnus may originate from the inferomedial border of the semitendinosus and the long head of the biceps femoris (Sigmon 1974) and/or from the inferomedial border of the quadratus femoris (our specimen VU PP1). Insertion is onto the medial ridge of the linea aspera on the dorsomedial surface of the femur, and onto the adductor tubercle and the medial femoral epicondyle (Beddard 1893, Boyer 1935b, Hepburn 1892, Sigmon 1974, Sonntag 1924a). The short head may also attach onto the lateral border of the common tendon of insertion of the adductor longus and adductor brevis (Beddard 1893) and/or onto a curved line from the mid-point to the intertrochanteric line to the intertrochanteric surface of the lesser trochanter (Boyer 1935b).
- Usual innervation: Short head by the obturator nerve (Hepburn 1892, Sigmon 1974); long head by flexores femoris nerve (Sigmon 1974), with the exception of two specimens where the nerve supply is derived from the nerve to the quadratus femoris and the sciatic nerve (to the hamstrings; Boyer 1935b);

according to Sigmon (1974) the flexor femoris nerve does not exist in modern humans, and the tibial nerve in apes (except in a few gorillas) does not supply the adductor magnus.

Adductor minimus (VU PP1 LSB = 6.21 g, RSB = 4.69 g)
- Usual attachments: The adductor minimus is absent from a third of great apes (Hepburn 1892, Sigmon 1974), being often described as a superior subdivision of the adductor magnus or as a deep slip of the adductor brevis (Hepburn 1892) or as the most inferior of the accessory adductor muscles (Appleton & Ghey 1929). When present, the adductor minimus runs from the inferior pubic ramus to the linea aspera on the dorsal femoral shaft (Hepburn 1892, Boyer 1935b; our dissections) and/or to the lateral lip of the accessory adductor groove (Appleton & Ghey 1929). In *Pongo* the muscle may be in two parts, with an additional insertion around the insertion of quadratus femoris, fusing with this muscle of the other side of the body (Boyer 1935b).
- Usual innervation: Data are not available.

Gracilis (VU PP1 LSB = 168.31 g, RSB = 167.08 g; Figs. 52, 59)
- Usual attachments: Origin is from the inferior pubic ramus near to the pubic symphysis and from the ischial ramus (Beddard 1893, Boyer 1935b, Ferrero 2011, Hepburn 1892, Sigmon 1974, Sonntag 1924a), sometimes extending to the whole pubic body (Hepburn 1892) and/or to the superior pubic ramus (Boyer 1935b, Robinson et al. 1972, Sonntag 1924a). Insertion is onto the anteromedial surface of the tibia (Boyer 1935b, Hepburn 1892, Sigmon 1974, Sonntag 1924a), and there is often an aponeurotic expansion to the fascia of the leg (Boyer 1935b, Hepburn 1892, Sonntag 1924a). The insertion of the gracilis is usually between those of the sartorius and semitendinosus, mediodistal to that of the sartorius, and lateroproximal to that of the semitendinosus (Beddard 1893, Boyer 1935b, Hepburn 1892, Owen 1830-1831, Sigmon 1974, Sonntag 1924a). One exception is found in our specimen GWU PP1, in which the insertion of the gracilis is laterodistal to that of the semitendinosus (Ferrero 2011, Ferrero et al. 2012).
- Usual innervation: Ventral ('anterior') branch of obturator nerve (Sigmon 1974).

Pectineus (VU PP1 LSB = 21.67 g, RSB = 13.74 g; Fig. 53)
- Usual attachments: Origin is from the superior pubic ramus (Boyer 1935b, Hepburn 1892, Sigmon 1974, Sonntag 1924a) and occasionally from the pubic tubercle and pubic body (our specimen VU PP1); the muscle may be fused with the adductor longus (Sigmon 1974). Insertion is onto the femur just distal to the lesser trochanter (Beddard 1893, Boyer 1935b, Hepburn 1892, Sigmon 1974, Sonntag 1924a). The groove of insertion is well-marked in *Gorilla* and *Hylobates*, and less so in *Pan* and *Pongo* (Appleton & Ghey 1929).

- Usual innervation: Femoral nerve and occasionally also by the ventral 'anterior' branch of the obturator nerve (Boyer 1935b, Sigmon 1974).
- Notes: To our knowledge, the **iliocapsularis** was not described as a distinct muscle in orangutans.

Biceps femoris (Caput longum VU PP1 LSB = 96.32 g, RSB = 90.81 g; Caput breve VU PP1 LSB = 41.7 g, RSB = 50.48 g; Figs. 50–51, 53)

- Usual attachments: The long head of the biceps femoris originates from the ischial tuberosity in common with semitendinosus (Beddard 1893, Boyer 1935b, Hamada 1985, Kumakura 1989, Owen 1830-1831, Prejzner-Morawska & Urbanowick 1971, Sigmon 1974, Sonntag 1924a) and often with semimembranosus (Beddard 1893), gluteus maximus (Sigmon 1974, Sonntag 1924a) and/or quadratus femoris (Beddard 1893). In our specimen GWU PP1 the long head was fused with the semitendinosus, semimembranosus and quadratus femoris muscles, while in our specimen VU PP1 it was with fused with the semitendinosus and semimembranosus only (Ferrero 2011, Ferrero et al. 2012). The short head of the biceps femoris originates from the dorsolateral femur in the region of the lateral ridge of the linea aspera (Beddard 1893, Ferrero 2011, Ferrero et al. 2012, Hamada 1985, Kumakura 1989, Owen 1830-1831, Prejzner-Morawska & Urbanowick 1971, Sigmon 1974, Sonntag 1924a), extending more distally in great apes than in *Hylobates* and *Homo* (Fick 1895ab, Sigmon 1974). It often also takes origin from the lateral intermuscular septum (Beddard 1893, Hamada 1985, Prejzner-Morawska & Urbanowick 1971). The long head of biceps femoris inserts onto the tibial head (Hepburn 1892, Prejzner-Morawska & Urbanowick 1971, Sigmon 1974), onto the tibial tuberosity or condyle (Boyer 1935b, Kaplan 1958a, Kaplan 1958b, Sigmon 1974, Sneath 1955, Sonntag 1924a), onto the fibular head and fascia of the leg (Beddard, 1893 Boyer, 1935b, Hamada 1985, Kumakura 1989, Owen 1830-1831, Prejzner-Morawska & Urbanowick 1971, Sigmon 1974), onto the iliotibial tract (Beddard 1893, Hamada 1985, Sigmon 1974), onto the capsule of the knee joint (Hepburn 1892, Prejzner-Morawska & Urbanowick 1971), and/or onto the distal femur and intermuscular septum (Boyer 1935b, Church 1861-1862, Prejzner-Morawska & Urbanowick 1971, Sigmon 1974, Sonntag 1924a). The short head inserts onto the fibular head and fascia of the leg (Beddard 1893, Boyer 1935b, Hamada 1985, Hepburn 1892, Kumakura 1989, Owen 1830-1831, Prejzner-Morawska & Urbanowick 1971, Sigmon 1974, Sonntag 1924a) and often onto the tibial tuberosity (Prejzner-Morawska & Urbanowick 1971, Sigmon 1974) and/or the lateral intermuscular septum (Boyer 1935b, Prejzner-Morawska & Urbanowick, 1971). The two heads of the muscle are fused in 4/9 orangutans according to Gibbs (1999). The long head may be bipartite in *Pongo* (Sigmon 1974); in this case the part inserting onto the distal femoral surface has been named the 'ischiofemoralis' (Prejzner-Morawska & Urbanowick 1971).

- Usual innervation: Caput longum by flexores femoris nerve as is usually the case in other apes (and not by the tibial nerve, which is usually the case in modern humans); caput breve by common fibular nerve as is usually the case in other apes and in modern humans (Sigmon 1974).

Semimembranosus (VU PP1 LSB = 113.07 g, RSB = 116.42 g; Figs. 53, 59)
- Usual attachments: From the ischial tuberosity, inferior and lateral to semitendinosus (Boyer 1935b, Ferrero 2011, Ferrero et al. 2012, Sigmon 1974) to the posterior surface of the medial tibial condyle (Beddard 1893, Boyer 1935b, Hepburn 1892, Sigmon 1974, Sonntag 1924a) and sometimes also to the tibial collateral ligament (Ferrero 2011, Ferrero et al. 2012). In a single *Pongo* specimen the semimembranosus and semitendinosus muscles were fused at their origins (Beddard 1893).
- Usual innervation: By the flexores femoris nerve, as is usually the case in other apes (not by the tibial nerve, as is usually the case in modern humans; Sigmon 1974).

Semitendinosus (VU PP1 LSB = 181.1 g, RSB = 188.86 g; Figs. 51–53, 59)
- Usual attachments: Origin is from the ischial tuberosity in common with the long head of biceps femoris (Boyer 1935b, Ferrero 2011, Ferrero et al. 2012, Hepburn 1892, Owen 1830-1831, Sigmon 1974, Sonntag 1924a) and often with the semimembranosus (Beddard 1893, Hepburn 1892, Owen 1830-1831, Sigmon 1974, Sonntag 1924a). Insertion is onto the tibial diaphysis (proximal medial surface), the tibial tuberosity (medial region), and onto the tibial collateral ligament, being distal to that of gracilis (Beddard 1893, Hepburn 1892, Sigmon 1974) although in some specimens it is medial (Boyer 1935b, Sigmon 1974) or above (Beddard 1893; our specimens GWU PP1 and VU PP1) that of this latter muscle. An oblique tendinous intersection in the muscle belly of semitendinosus is occasionally present (Hepburn 1892, MacAlister 1871). In some specimens there is, in addition, an aponeurotic expansion to the fascia of the leg in this region (Boyer 1935b, Owen 1830-1831, Sonntag 1924a). The insertion of the semitendinosus in *Pongo* usually extends further distally than in *Homo* and *Gorilla* (Hepburn 1892, Owen 1830-1831).
- Usual innervation: By the flexores femoris nerve, as is usually the case in other apes (not by the tibial nerve, as is usually the case in modern humans; Sigmon 1974).

Extensor digitorum longus (VU PP1 LSB = 46.97 g, RSB = 49.51 g; Figs. 52, 56)
- Usual attachments: Origin is from the head and medial crest of the fibula, and from the intermuscular septum (Beddard 1893, Boyer 1935b, Kaplan 1958a, Lewis 1966, Sonntag 1924a) and often also from the lateral tibial condyle (Beddard 1893, Boyer 1935b, Lewis 1966) and/or occasionallly from the interosseous membrane (Beddard 1893). The origin of the muscle is sometimes (e.g., our specimens GWU PP1 and VU PP1) fused with that of

the tibialis anterior. Insertion is usually onto the dorsal aponeurosis of digits 2–5 (Beddard 1893, Boyer 1935b, Hepburn 1892, Lewis 1966, Sonntag 1924a) but in a few cases there is no insertion to digit 2 (Beddard 1893, Owen 1830-1831). In our VU PP1 the most independent tendon was that inserting onto the digit 5; the tendons inserting onto digits 2 and 3 were closely associated, and the tendon inserting onto digit 4 took its origin from the tendon to digits 3 (Ferrero 2011, Ferrero et al. 2012). In our specimen GWU PP1 the tendon to digit 2 was tiny and fused with the tendon to of the extensor digitorum brevis to digit 2, both tendons inserting together onto the dorsal aponeurosis of this digit (Ferrero 2011, Ferrero et al. 2012).
- Usual innervation: Data are not available.

Extensor hallucis longus (VU PP1 LSB = 8.6 g, RSB = 8.06 g; Figs. 52, 56)
- Usual attachments: From the medial surface of the fibula (Beddard 1893, Lewis 1966, Sonntag 1924a) and often from the interosseous membrane (Beddard 1893, Boyer 1935b, Lewis 1966) to the terminal phalanx of the hallux (Beddard 1893, Boyer 1935b, Ferrero 2011, Ferrero et al. 2012, Lewis 1966, Owen 1830-1831, Sonntag, 1924a) and often to the shaft of metatarsal I (Beddard 1893, Boyer 1935b, Hepburn 1892). In our specimen GWU PP1 the origin was partially fused with that of the extensor digitorum longus.
- Usual innervation: Data are not available.
- Notes: To our knowledge the **fibularis tertius** ('peroneus tertius') was never found in orangutans.

Tibialis anterior (VU PP1 LSB = 67.46 g, RSB = 71.59 g; Fig. 52)
- Usual attachments: Origin is from the lateral tibia (Beddard 1893, Boyer 1935b, Ferrero 2011, Ferrero et al. 2012, Lewis 1966, Owen 1830-1831, Sonntag 1924a) and occasionally from the interosseous membrane (Boyer 1935b) and/or the extensor digitorum longus (Sonntag 1924a; our specimens GWU PP1 and VU PP1). Insertion is onto the plantar surface of the medial cuneiform (Beddard 1893, Boyer 1935b, Ferrero 2011, Ferrero et al. 2012, Hepburn 1892, Lewis 1966, Sonntag 1924a) and the proximal end of metatarsal I (Boyer 1935b, Hepburn 1892, Lewis 1966, Owen 1830-1831, Sonntag 1924a).
- Usual innervation: Data are not available.

Fibularis brevis (VU PP1 LSB = 25.00 g, RSB = 24.53 g; Figs. 52, 54)
- Usual attachments: From the fibular distal lateral surface and intermuscular septa (Boyer 1935b, Owen 1830-1831, Sonntag 1924a) and sometimes also from the dorsal and ventral fibular surfaces (Owen 1830-1831, Sonntag 1924a) and/or the fascia of the leg (Beddard 1893), to the tuberosity at the base of metatarsal V (Beddard 1893, Boyer 1935b, Hepburn 1892, Lewis 1966, Owen 1830-1831, Sonntag 1924a) and often to the extensor digitorum brevis tendon to digit 5 (Beddard 1893, Ferrero 2011, Ferrero et al. 2012, Hepburn 1892, Sonntag 1924a) and/or occasionally onto connective tissue (Boyer 1935b). The

tibialis anterior is sometimes fused with the extensor digitorum longus and fibularis longus (Owen 1830-1831, Sonntag 1924a; our specimen GWU PP1) and/or connected with the flexor hallucis longus (Owen 1830-1831).

- Usual innervation: Data are not available.
- Synonymy: Peroneus brevis (Gibbs 1999).

Fibularis longus (VU PP1 LSB = 38.97 g, RSB = 40.42 g; Figs. 52, 54, 58)

- Usual attachments: From the fibular head and proximal fibula (Beddard 1893, Boyer 1935b, Lewis 1966, Owen 1830-1831, Ruge 1878a, Sonntag 1924a) and occasionally from fascia (Boyer 1935b), passing in a groove on the cuboid bone (Lewis 1966) and inserting onto the tuberosity of metatarsal I (Beddard 1893, Boyer 1935b, Ferrero 2011, Ferrero et al. 2012, Hepburn 1892, Lewis 1966, Owen 1830-1831, Sonntag 1924a). In the tendon of the fibularis longus a slight thickening, or nodule, has been reported in the great apes (e.g., Manners-Smith 1908). The fibularis longus may be fused with the fibularis brevis and extensor digitorum longus (Beddard 1893, Owen 1830-1831, Sonntag 1924a) and/or connected with the flexor hallucis longus (Beddard 1893, Owen 1830-1831, Sonntag 1924a).
- Usual innervation: Data are not available.
- Synonymy: Peroneus longus (Gibbs 1999).

Gastrocnemius (Caput laterale VU PP1 LSB = 45.8 g, RSB = 60.68 g; Caput mediale VU PP1 LSB = 67.47 g, RSB = 79.66 g; Figs. 54, 60)

- Usual attachments: From the medial and lateral femoral condyles and the capsule of the knee joint (Beddard 1893, Boyer 1935b, Frey 1913, Hepburn 1892, Sonntag 1924a) to the calcaneal tuberosity (Beddard 1893, Boyer 1935b, Ferrero 2011, Sonntag 1924a). The gastrocnemius joins with soleus and its lateral head may have a common origin with the long head of flexor hallucis longus (Beddard 1893, Boyer 1935b). In our GWU PP1 specimen the lateral head of the gastrocnemius had a common origin with that of the plantaris. The calcaneal tendon in *Hylobates* and *Pongo* it clearly more defined and stronger than in *Gorilla* and *Pan*, although not as defined as in *Homo* (Gibbs 1999).
- Usual innervation: Data are not available.

Plantaris (Fig. 60)

- Usual attachments: The plantaris has been reported in two orangutans (Primrose 1899, Sandifort 1840), and was present in our GWU PP1 specimen, being a penniform muscle, extremely long and flat. When present, the plantaris runs from the lateral tibial condyle to the calcaneal posterior tuberosity (Ferrero 2011, Ferrero et al. 2012). The origin of the plantaris in GWU PP1 is divided into two bellies; both of them take origin from the lateral tibial condyle but the most lateral head is fused with the flexor hallucis longus and the lateral head of gastrocnemius; the most medial head is fused distally with the flexor hallucis longus.
- Usual innervation: Data are not available.

Soleus (VU PP1 LSB = 96.75 g, LSB = 101.11 g; Figs. 54, 55, 60)

- Usual attachments: Origin is said to be usually from the head and superodorsal aspect of the fibular shaft (Beddard 1893, Boyer 1935b, Hepburn 1892, Huxley 1864, Lewis 1962, Sonntag 1924a) but in our specimens GWU PP1 and VU PP1 it was exclusively from the fibular head (Ferrero 2011, Ferrero et al. 2012). The tibial origin is absent from *Pongo* (Boyer 1935b, Church 1861-1862, Owen 1830-1831), and may be included with the lateral part of the gastrocnemius according to Michaëlis (1903). Insertion is onto the calcaneal tuberosity (Beddard 1893, Boyer 1935b, Sonntag, 1924a). In our specimens GWU PP1 and VU PP1 there was an additional muscular belly of the soleus, i.e., a third belly located deeper and laterally to the main body of the soleus. This additional belly originated from the intermuscular septum by a short and flat tendon and inserted onto the calcaneal tuberosity. This additional soleus muscular belly is described as present in about 3% of modern humans ('**accessory soleus**' sensu Moore & Dalley 1999).
- Usual innervation: Data are not available.

Flexor digitorum longus (VU PP1 LSB = 84.21 g, LSB = 63.44 g; Figs. 54–55, 58, 60–61)

- Usual attachments: It arises from the posterior aspect of the tibial shaft (Beddard 1893, Sonntag 1924a), its origin being usually more extensive than in other apes (Beddard 1893, Boyer 1935b), and sometimes also arises from the tibialis posterior fascia (Boyer 1935b). The most common insertion appears to be to digits 2, 4 and 5 (Gibbs 1999), onto the distal phalanx of the digits. Digit 5 is seemingly supplied in all orangutans, digit 4 is in 12 out of 22 orangutans, and digit 2 in 15 out of 22 orangutans according to Gibbs's (1999) review. The muscle may be fused with the flexor digitorum brevis (Boyer 1935b).
- Usual innervation: Data are not available.
- Synonymy: Flexor digitorum medialis (Gibbs 1999).

Flexor hallucis longus (VU PP1 LSB = 64.53 g, LSB = 85.35 g; Figs. 54–55, 58, 60)

- Usual attachments: Origin is from the interosseous membrane, the posterior crural intermuscular septum (Beddard 1893, Boyer 1935b) and/or the distal fibula (Beddard 1893, Boyer 1935b, Owen 1830-1831, Sonntag 1924a). The muscle often has a femoral head and a fibular head, the femoral head being often fused with gastrocnemius (Boyer 1935b, Ferrero 2011, Ferrero et al. 2012, Owen 1830-1831, Sonntag 1924a); in our specimen GWU PP1 the femoral head had an additional origin from the tibial collateral ligament. An origin from the lateral femoral condyle is also frequently present in *Pongo* (Beddard 1893, Boyer 1935b, Hepburn 1892, Owen 1830-1831, Sonntag 1924a). In the space usually occupied by the flexor hallucis longus there may be a small muscle arising from metatarsal I and inserting onto the hallux (Owen 1830-1831).

Insertion of the flexor hallucis longus is onto the base of the distal phalanx of digits 3 and 4 (Beddard 1893, Boyer 1935b, Chapman 1880, Hepburn 1892, Keith 1894a, Lewis 1962, Michaëlis 1903, Owen 1830-1831, Sonntag 1924a). No insertion onto digit 1 was reported for *Pongo*.
- Usual innervation: Data are not available.
- Synonymy: Flexor digitorum lateralis (Gibbs 1999).

Popliteus (VU PP1 LSB = 27.59 g, LSB = 27.51 g; Figs. 55, 60)
- Usual attachments: One muscle belly of the popliteus often takes origin from the lateral femoral epicondyle (Beddard 1893, Boyer 1935b, Hepburn 1892, Sonntag 1924a). The second head of origin (the only one present in our specimen GWU PP1) originates from the fibular head and adjacent knee joint capsule (Hepburn 1892). In our VU PP1 specimen the popliteus muscle also originates from the popliteal oblique ligament (and sesamoid bone: see below) and from the lateral meniscus (Ferrero 2011, Ferrero et al. 2012). Insertion of the popliteus is onto the posterior tibial surface (Beddard 1893, Boyer 1935b, Ferrero 2011, Ferrero et al. 2012, Sonntag 1924a) and occasionally onto the tibial proximal lateral and medial surface and/or onto the tibial collateral ligament (Gibbs 1999). A sesamoid bone is present in the tendon of the lateral tibial condyle, in 18/21 *Pongo* (Gibbs 1999).
- Usual innervation: Tibial nerve (Hepburn 1892).

Tibialis posterior (VU PP1 LSB = 28.51 g, LSB = 32.25 g; Figs. 55, 59–60)
- Usual attachments: From the interosseous membrane, tibia and fibula (Beddard 1893, Boyer 1935b), to the navicular bone (Hepburn 1892, Lewis 1964) and often the sheath of the tendon of fibularis longus (Boyer 1935b, Hepburn 1892, Lewis 1964, Sonntag 1924a) and/or occasionally the medial/intermediate cuneiforms (Boyer 1935b, Lewis 1964). The sesamoid fibrocartilage present in the tendon of tibialis posterior of *Homo* is absent from *Pongo* (Owen, 1830-1831).
- Usual innervation: Data are not available.

Extensor digitorum brevis (VU PP1 LSB = 25.03 g, LSB = 29.68 g; Figs. 54, 56)
- Usual attachments: From the calcaneus (Beddard 1893, Boyer 1935b, Hepburn 1892, Lewis 1966, Owen 1830-1831, Ruge 1878a, Sonntag 1924a) and occasionally from the base of metatarsal III (our specimen VU PP1), to the dorsal aponeurosis of digits 2–4 (Beddard 1893, Boyer 1935b, Hepburn 1892, Lewis 1966, Owen 1830-1831, Ruge 1878a, Sonntag 1924a). One specimen of *Pongo* had a double insertion onto digit 2, one tendon to the proximal phalanx and the other tendon to the middle and distal phalanges of this digit (Owen 1830-1831).
- Usual innervation: Data are not available.

Extensor hallucis brevis (VU PP1 LSB = 3.67 g, LSB = 3.56 g; Figs. 52, 56)
- Usual attachments: From the calcaneus and navicular bone (Owen 1830-1831, Ruge 1878a) and often also from the extensor digitorum brevis (Beddard 1893, Boyer 1935b, Hepburn 1892, Lewis 1966, Sonntag 1924a), to the dorsal aspect of the base of the proximal phalanx of the hallux (Boyer 1935b, Hepburn 1892, Lewis 1966, Owen 1830-1831, Sonntag 1924a). The extensor hallucis brevis is usually a separate muscle in *Pongo* (Beddard 1893, Boyer 1935b, Hepburn 1892, Lewis 1966, Sonntag 1924a) but it may be fused with the tendon of the extensor hallucis longus (Boyer 1935b).
- Usual innervation: Data are not available.

Abductor digiti minimi (VU PP1 LSB = 14.64 g, LSB = 14.63 g; Figs. 54, 56-57)
- Usual attachments: From the medial and lateral margins of the calcaneus (Beddard 1893, Boyer 1935b, Sonntag 1924a) and often from the plantar aponeurosis (Beddard 1893, Boyer 1935b) and/or the sheath of fibularis longus (Sonntag 1924a), to the proximal phalanx of digit 5 (Beddard 1893, Sonntag 1924a) and often the base of metatarsal V (Beddard 1893, Boyer 1935b, two tendons of insertion to this bone in our GWU PP1 specimen), and occasionally the dorsal expansion of the extensor tendons (Boyer 1935b; our specimen VU PP1).
- Usual innervation: Data are not available.
- Synonymy: Abductor digiti quinti (Gibbs 1999).

Abductor hallucis (VU PP1 LSB = 4.69 g, LSB = 4.49 g; Figs. 57, 61)
- Usual attachments: The abductor hallucis runs from the medial and plantar surfaces of the calcaneus (Beddard 1893, Boyer 1935b, Brooks 1887, Hepburn 1892, Sonntag 1924a) and sometimes also from the medial part of the plantar aponeurosis (Beddard 1893, Church 1861-1862, Hepburn 1892) and/or the 'posterior' surface of the calcaneus (Boyer 1935b), to the base of the proximal phalanx of the hallux (Beddard 1893, Boyer 1935b, Brooks 1887, Hepburn 1892, Owen 1830-1831, Sonntag 1924a). In some orangutans the abductor hallucis is laterally blended with the flexor digitorum brevis (Sonntag 1924a; our specimen VU PP1) and/or with the flexor hallucis brevis (Beddard 1893, Brooks 1887, Ruge 1878b).
- Usual innervation: Medial plantar nerve (Brooks 1887).

Abductor os metatarsi digiti minimi (VU PP1 LSB = 2.46 g, LSB = 3.2 g; Fig. 62)
- Usual attachments: Some orangutans have a separate slip that goes to the base of MI (Brooks 1887, Church 1861-1862, Hepburn 1892, Sonntag 1924a), which has been named the '**abductor ossis metacarpi hallucis**' ('**abductor os metatarsi digiti quinti**' *sensu* Gibbs 1999; abductor os metatarsi digiti minimi *sensu* the present work). In our specimen VU PP1 there is an abductor os metatarsi digiti minimi (Fig. 62). When present, this muscle is located immediately overlying

the inner head of abductor digiti minimi, deep to the abductor digiti minimi and lateral to the flexor digiti minimi brevis, and runs from the plantar surface of the calcaneus in common with the lateral part of flexor digitorum brevis, to the lateral and plantar aspects of the base of metatarsal V (Primrose 1899).

- Usual innervation: Data are not available.

Flexor digitorum brevis (VU PP1 LSB = 29.23 g, LSB = 31.06 g; Figs. 57, 61)

- Usual attachments: Origin is from the calcaneus, usually from its medial and plantar surfaces (Beddard 1893, Boyer 1935b, Hepburn 1892, Sarmiento 1983, Sonntag 1924a). According to Boyer (1935b) the superficial head may also originate from the distal plantar aponeurosis, calcaneal tendon and medial soleus; the latter two origins were not found in the two specimens where we dissected the foot in detail (GWU PP1 and VU PP1). Insertion is usually onto the middle phalanx of digits 2 and 3 (Beddard 1893, Boyer 1935b, Chapman 1880, Hepburn 1892, Owen 1830-1831, Primrose 1900, Sarmiento 1983, Sonntag 1924a) and often also 4 (Beddard 1893, Boyer 1935b, Ferrero 2011, Ferrero et al. 2012, Hepburn 1892, Primrose 1900, Sarmiento 1983). When present, the tendon to digit 4 is derived from a deep head (Boyer 1935b, Hepburn 1892, Primrose 1900, Sarmiento 1983), and this head occasionally supplies digit IV (Hepburn, 1892, Sarmiento, 1983). Sometimes there is a tendon to digit 5, which is fused with the flexor digitorum longus, and a tendon from the deep head may join the fourth lumbrical muscle proximally (Boyer 1935b). the flexor digitorum brevis may be fused with the abductor hallucis (Sonntag 1924a).
- Usual innervation: Data are not available.

Quadratus plantae

- Usual attachments: The lateral head originates from the lateral margin of the plantar surface of the calcaneus (Chapman 1880, Hepburn 1892); a medial head is sometimes present in *Pongo* (e.g., Lewis 1962). The quadratus plantae usually inserts onto the flexor digitorum longus, particularly onto its tendon to digit 5 (Church 1861-1862). In the specimens where we dissected the foot in detail (GWU PP1 and VU PP1) this muscle was however absent.
- Usual innervation: Data are not available.
- Synonymy: Flexor accessorius (Gibbs 1999).

Lumbricales (VU lumbricales 1, 2, 3 and 4 PP1 LSB = 9.22 g, LSB = 10.53 g; Fig. 58)

- Usual attachments: The first lumbrical has a single head from the flexor digitorum longus tendon to digit 2 (Beddard 1893, Boyer 1935b, Chapman 1880, Hepburn 1892, Sonntag 1924a). The second lumbrical has a double origin (Beddard 1893, Boyer 1935b, Chapman 1880, Hepburn 1892, Sonntag 1924a), as observed in our specimen GWU PP1, from the tendons of the flexor hallucis longus and flexor digitorum longus. However, as described by Boyer (1935b), in our specimen VU PP1 the left second lumbrical had a single origin from the tendon of the flexor digitorum longus to digit 3. The third lumbrical has a double origin in some specimens (Beddard 1893, Boyer 1935b; our specimens GWU PP1 and VU PP1), i.e., from the tendons of the flexor hallucis longus and the tendon of the flexor digitorum longus to digit 4, but in some cases it may have a single head (Hepburn 1892, Sonntag 1924a). The fourth lumbrical has a double origin in some specimens (Primrose 1899; our specimens GWU PP1 and VU PP1), both from the tendon of the flexor digitorum longus to digit 5, but in many orangutans it has instead a single head of origin (Boyer 1935b, Hepburn 1892, Sonntag 1924a). The tendons of the lumbricales usually insert directly onto the medial side of the base of the proximal phalanx of digits 2, 3, 4 and 5 and also radiate onto the extensor aponeurosis of these digits (Boyer 1935b).

- Usual innervation: Data are not available.

Adductor hallucis (Caput obliquum VU PP1 LSB = 17.89 g, LSB = 7.94 g; Caput transversum VU PP1 LSB = 10.09 g, LSB = 11.15 g; Figs. 52, 61, 56–58)

- Usual attachments: The oblique head originates from metatarsals II and III (Beddard 1893, Brooks 1887) and often also from the sheath of the tendon of the fibularis longus (Beddard 1893, Brooks 1887, Sonntag 1924a). The transverse head takes origin from the plantar surface of metatarsals II and III (Beddard 1893, Boyer 1935b, Brooks 1887) and often from the interosseous fascia (Boyer 1935b, Brooks 1887), from a slip associated with metatarsal II (Boyer 1935b), from a fascial band extending from metatarsal II to the third tarsometatarsal joint (Brooks 1887), and/or from connective tissue (Beddard 1893, Boyer 1935b). The oblique head inserts onto the base of the proximal phalanx of the hallux (Brooks 1887) and often also onto metatarsal I (Beddard 1893, Boyer 1935b, Brooks 1887) and/or onto a sesamoid bone which is situated at the combined oblique head/flexor hallucis brevis insertion (Beddard 1893, Brooks 1887, Ruge 1878b, our specimens GWU PP1 and VU PP1). The insertion of the transverse head is onto the base of the proximal phalanx (Boyer 1935b, Brooks 1887, Owen 1830-1831) and often also onto the base of metatarsal I (Beddard 1893, Brooks 1887); in our specimen VU PP1 the transverse head was divided into two bellies. Some fibres of the adductor hallucis that insert onto metatarsal I may form a somewhat separated '**adductor opponens**' (Brooks 1887, Church 1861-1862).
- Usual innervation: Deep branch of lateral plantar nerve (Brooks 1887, Hepburn 1892), although an additional innervation from the medial plantar nerve has been also found (Brooks 1887, Hepburn 1892, Sonntag 1924a).
- Notes: In modern humans the '**transversalis pedis**' (not listed in Terminologia Anatomica 1998) corresponds to the caput transversum of the adductor hallucis.

Flexor digiti minimi brevis (VU PP1 LSB = 4.24 g, LSB = 7.23 g; Figs. 58, 61–62)
- Usual attachments: From the base of metatarsal V (Boyer 1935b, Sonntag 1924a) and sometimes also from the sheath of the tendon of fibularis longus (Sonntag 1924a), from the plantar aponeurosis (Boyer 1935b) and/or from the plantar tarsometatarsal ligament (Ferrero 2011, Ferrero et al. 2012), to the medial region of the base of the proximal phalanx of digit 5 (Boyer 1935b, our dissections).
- Synonymy: Flexor digiti quinti brevis or flexor digiti minimi (Gibbs 1999).

Flexor hallucis brevis (Caput mediale VU PP1 LSB = 5.75 g, LSB = 9.21 g; Caput laterale VU PP1 LSB = 5.41 g, LSB = 12.27 g; Fig. 57)
- Usual attachments: From the intermediate cuneiform and the tendon of tibialis posterior (Boyer 1935b, Lewis 1964) and often also from the medial and lateral cuneiforms (Brooks 1887, Church 1861-1862) and/or from the navicular bone and the plantar fascia (Church 1861-1862), to the proximal phalanx of the hallux (Beddard 1893, Boyer 1935b, Brooks 1887, Church 1861-1862, Owen 1830-1831) and often also the distal region of metatarsal I (Beddard 1893, Brooks 1887, Church 1861-1862). The medial and lateral heads are at least sometimes separated by a septum (Brooks 1887). The medial head of the muscle is sometimes fused with the abductor hallucis (Beddard 1893, Bischoff 1870, Ruge 1878b), and the lateral head with the opponens hallucis (Boyer 1935b).
- Usual innervation: Medial plantar nerve (Brooks 1887).

Opponens digiti minimi (VU PP1 LSB = 1.14 g, LSB = 0.35 g; Fig. 62)
- Usual attachments: The opponens digiti minimi is often described as part of the flexor digiti minimi and is usually considered to be missing as a separate muscle in *Pongo* (Boyer 1935b, Fick 1895ab, Hepburn 1892, Sonntag 1924a). However it was clearly present as a distinct muscle in the two specimens where we dissected the foot in detail (GWU PP1 and VU PP1). When present, it runs from the plantar aponeurosis and the tendon sheath of the fibularis longus (Boyer 1935b, Ferrero 2011, Ferrero et al. 2012, Primrose 1899) to the lateral surface of metatarsal V (Primrose 1899; our specimen GWU PP1) and/or the proximal phalanx of digit 5 (our specimen VU PP1).
- Usual innervation: Data are not available.
- Synonymy: Opponens digiti quinti (Gibbs 1999).

Opponens hallucis
- Usual attachments: Within apes, this muscle has been described in detail only in *Pongo* (Bischoff 1870, Boyer 1935b, Brooks 1887, Chapman 1880, Church 1861-1862, Huxley 1864, Primrose 1899. It often originates from the medial cuneiform (Church 1861-1862), passes beneath the abductor hallucis, and is entirely or partially fused with the lateral belly of flexor hallucis brevis (Boyer 1935b, Brooks 1887, Primrose 1899). In a single *Pongo*, where the muscle was

fused with flexor hallucis brevis (Brooks 1887), the muscle originated from a cartilaginous nodule in the tendon of tibialis posterior. Insertion of the opponens hallucis is onto the distal, middle third or entire lateral length of metatarsal I (Boyer 1935b, Brooks 1887, Church 1861-1862, Huxley 1864, Primrose 1899) and sometimes also onto the base of the proximal phalanx of the hallux (Boyer 1935b). It should however be noted that in the two specimens where we dissected the foot in detail (GWU PP1 and VU PP1) the opponens hallucis was not present as a distinct muscle.

- Usual innervation: Medial plantar nerve (Brooks 1887).

Interossei dorsales (Interossei dorsales 1, 2, 3 and 4 VU PP1 LSB = 36.78 g, LSB = 38.21 g; Figs. 56–58, 61)

- Usual attachments: There are usually four muscles, their reference line for action being through digit 3 (Boyer 1935b, Duvernoy 1855-1856). The first dorsal interosseous originates from the medial cuneiform and the medial side of metatarsal II (Boyer 1935b, Brooks 1887, Sonntag 1924a) and occasionally also from the metatarsal I (Ferrero 2011, Ferrero et al. 2012). The second dorsal interosseous originates from the lateral side of metatarsal II and the medial side of metatarsal III (Boyer 1935b). The third dorsal interosseous originates from the lateral side of metatarsal III and the medial side of metatarsal IV (Boyer 1935b). The fourth dorsal interosseous originates from the lateral side of metatarsal IV and medial side of metatarsal V (Boyer 1935b). The first dorsal interosseous muscle inserts onto the medial side of the proximal phalanx of digit 2 (Boyer 1935b, Owen 1830-1831). The second dorsal interosseous inserts onto the medial side of the proximal phalanx of digit 3 (Boyer 1935b, Brooks 1887). The third dorsal interosseous inserts onto the lateral side of the proximal phalanx of digit 3 (Boyer 1935b, Brooks 1887). The fourth dorsal interosseous inserts onto the lateral side of the proximal phalanx of digit 4 (Boyer 1935b, Brooks 1887).
- Usual innervation: Data are not available.

Interossei plantares (Interossei plantares 1, 2 and 3 VU PP1 LSB = 14.56, LSB = 20.1 g; Fig. 58)

- Usual attachments: There are usually three muscles, their reference line for action being digit 3 (Boyer 1935b, Sonntag 1924a). The first plantar interosseous usually originates from metatarsal II (Boyer 1935b). On the left foot of our specimen VU PP1 there were however four origins: from the lateral aspect of the base of metatarsal I, from the distomedial portion of the navicular bone, from the base of metatarsal II and from the medial aspect of the base of metatarsal III. On the right foot of this specimen VU PP1 the muscle had three origins, i.e., the ones described just above except the one from the lateral aspect of the base of metatarsal I (Ferrero 2011, Ferrero et al. 2012). The second plantar interosseous usually originates, in *Pongo*, from the medial side of metatarsal IV

(Boyer 1935b); in our VU PP1 specimen there were however two origins: from the lateral aspect of the base of metatarsal III and from the medial aspect of metatarsal IV (Ferrero 2011, Ferrero et al. 2012). In *Pongo*, the third plantar interosseous usually originates from the medial side of metatarsal V (Boyer 1935b, Ferrero 2011, Ferrero et al. 2012). The first plantar interosseous usually inserts onto the lateral side of the proximal phalanx of digit 2 (Boyer 1935b) and/or onto the medial aspect of the base of the proximal phalanx of this digit (Ferrero 2011, Ferrero et al. 2012). The second plantar interosseous inserts onto the lateral side of the proximal phalanx of digit 4 (Boyer 1935b). The third plantar interosseous inserts onto the lateral side of the proximal phalanx of digit 5 (Boyer 1935b).

- Usual innervation: Data are not available.

Appendix I
Literature Including Information about the Muscles of Orangutans

Aiello L, Dean C (1990) *An introduction to human evolutionary anatomy*. San Diego: Academic Press.

Anderton JC (1988) Anomalies and atavisms in appendicular myology. In *Orang-utan Biology* (ed. Schwartz JH), pp. 331–345. New York: Oxford University Press.

Andrews P, Groves CP (1976) Gibbons and brachiation. In *Gibbon and Siamang, Vol. 4* (ed. Rumbaugh DM), pp. 167–218. Basel: Karger.

Appleton AB, Ghey PHR (1929) An example of the cervico-costo-humeral muscle of Gruber. *J Anat* 63, 434–436.

Ashton EH, Oxnard CE (1963) The musculature of the primate shoulder. *Trans Zool Soc Lond* 29, 553–650.

Aziz MA (1980) Anatomical defects in a case of trisomy 13 with a D/D translocation. *Teratology* 22, 217–227.

Aziz MA (1981) Possible "atavistic" structures in human aneuploids. *Am J Phys Anthropol* 54, 347–53.

Aziz MA, Dunlap SS (1986) The human extensor digitorum profundus muscle with comments on the evolution of the primate hand. *Primates* 27, 293–319.

Barnard WS (1875) Observations on the membral musculation of *Simia satyrus* (Orang) and the comparative myology of man and the apes. *Proc Amer Assoc Adv Sci* 24, 112–144.

Beddard FE (1893) Contributions to the anatomy of the anthropoid apes. *Trans Zool Soc Lond* 13, 177–218.

Bischoff TLW (1870) Beitrage zur Anatomie des *Hylobates leuciscus* and zueiner vergleichenden Anatomie der Muskeln der Affen und des Menschen. *Abh Bayer Akad Wiss Münchhen Math Phys Kl* 10, 197–297.

Blake ML (1976) The queantitative mycology of the hind limb of Primates with special reference to their locomotor adaptations. PhD Thesis, Magdalene College, University of Cambridge.

Bluntschli H (1929) Die kaumuskulatur des Orang-Utan und ihre Bedeutung für die Formung des Schädels. *Gegen Morphol Jahrb* 63, 531–606.

Bojsen-Møller F (1978) Extensor carpi radialis longus muscle and the evolution of the first intermetacarpal ligament. *Am J Phys Anthropol* 48, 177–184.

Bolk L (1902) Beiträge zur Affenanatomie, III, Der Plexus cervico-brachialis der Primaten. *Petrus Campter* 1, 371–566.

Boyer EL (1935a) The cranio-mandibular musculature of the orang-utan, *Simia satyrus*. *Am J Phys Anthropol* 24, 417–427.

Boyer EL (1935b) The musculature of the inferior extremity of the orang-utan, *Simia satyrus. Am J Anat* 56, 192–256.

Brandes G (1932) Über den Kehlkopf des Orang-Utans in verschiedenen Altersstadien mit besonderer Berücksichtigung der Kehlsackfrage. *Gegen Morphol Jahrb* 69, 1–61.

Broca P (1869) L'ordre des primates—parallele anatomique de l'homme et des singes. *Bull Soc Anthropol Paris* 4, 228–401.

Brooks HSJ (1886a) On the morphology of the intrinsic muscles of the little finger, with some observations on the ulnar head of the short flexor of the thumb. *J Anat Physiol* 20, 644–661.

Brooks HSJ (1886b) Variations in the nerve supply of the flexor brevis pollicis muscle. *J Anat Physiol* 20, 641–644.

Brooks HSJ (1887) On the short muscles of the pollex and hallux of the anthropoid apes, with special reference to the opponens hallucis. *J Anat Physiol* 22, 78–95.

Brown B (1983) An evaluation of primate caudal musculature in the identification of the ischiofemoralis muscle. *Am J Phys Antropol* 60, 177–178.

Brown B, Ward SC (1988) Basicranial and facial topography in *Pongo* and *Sivapithecus*. In *Orang-Utan Biology* (ed. Schwartz JH), pp. 247–260. Oxford: Oxford University Press.

Cachel S (1984) Growth and allometry in primate masticatory muscles. *Arch Oral Biol* 29, 287–293.

Cave AJE (1979) The mammalian temporo-pterygoid ligament. *J Zool Lond* 188, 517–532.

Chapman HC (1880) On the structure of the orang outang. *Proc Acad Nat Sci Philad* 32, 160–175.

Church WS (1861-1862) On the myology of the orang utang (*Simia morio*). *Natl Hist Rev* 1, 510–516; 2, 82–94.

Day MH, Napier J (1961) The two heads of the flexor pollicis brevis. *J Anat* 95, 123–130.

Day MH, Napier J (1963) The functional significance of the deep head of flexor pollicis brevis in primates. *Folia Primatol* 1, 122–134.

Dean MC (1984) Comparative myology of the hominoid cranial base-I—the muscular relationships and bony attachments of the digastric muscle. *Folia Primatol* 43, 234–48.

Dean MC (1985) Comparative myology of the hominoid cranial base-II—the muscles of the prevertebral and upper pharyngeal region. *Folia Primatol* 44, 40–51.

Deniker J, Boulart R (1885) Les sacs laryngiens, les excroissances adipeuses, les poumons, le cerveau, etc. des orang-utans. *Nouv Arch Mus Hist Nat, ser.* 3, 7, 35–56.

Diogo R, Abdala V (2010) Muscles of vertebrates—comparative anatomy, evolution, homologies and development. Oxford: Taylor & Francis.

Diogo R, Wood BA (2008) Comparative anatomy, phylogeny and evolution of the head and neck musculature of hominids: a new insight. *Am J Phys Anthropol, Suppl* 46, 90.

Diogo R, Wood BA (2009) Comparative anatomy and evolution of the pectoral and forelimb musculature of primates: a new insight. *Am J Phys Anthropol, Meeting Suppl* 48, 119.

Diogo R, Wood BA (2011) Soft-tissue anatomy of the primates: phylogenetic analyses based on the muscles of the head, neck, pectoral region and upper limb, with notes on the evolution of these muscles. *J Anat* 219, 273–359.

Diogo R, Wood BA (2012) Comparative anatomy and phylogeny of primate muscles and human evolution. Oxford: Taylor & Francis.

Diogo R, Abdala V, Lonergan N, Wood BA (2008) From fish to modern humans—comparative anatomy, homologies and evolution of the head and neck musculature. *J Anat* 213, 391–424.

Diogo R, Abdala V, Aziz MA, Lonergan N, Wood BA (2009a) From fish to modern humans—comparative anatomy, homologies and evolution of the pectoral and forelimb musculature. *J Anat* 214, 694–716.

Diogo R, Wood BA, Aziz MA, Burrows A (2009b) On the origin, homologies and evolution of primate facial muscles, with a particular focus on hominoids and a suggested unifying nomenclature for the facial muscles of the Mammalia. *J Anat* 215, 300–319.

Diogo R, Richmond BG, Wood B (2012) Evolution and homologies of modern human hand and forearm muscles, with notes on thumb movements and tool use. *J Hum Evol* 63, 64–78.

Duckworth WLH (1904) *Studies from the Anthropological Laboratory, the Anatomy School*, Cambridge, London: C. J. Clay & Sons.

Duckworth WLH (1912) On some points in the anatomy of the plica cocalis. *J Anat Physiol* 47, 80–115.

Duckworth WLH (1915) *Morphology and anthropology (2nd ed.)*. Cambridge: Cambridge University Press.

Duvernoy M (1855-1856) Des caracteres anatomiques de grands singes pseudoanthropomorphes anthropomorphes. *Arch Mus Natl Hist Nat Paris* 8, 1–248.

Edgeworth FH (1935) *The cranial muscles of vertebrates*. Cambridge: Cambridge University Press.

Eggeling H (1896) Zur Morphologie der Darnm-muskulatur. *Morphol Jahrb* 24, 511–631.

Elftman HO (1932) The evolution of the pelvic floor of primates. *Am J Anat* 51, 307–346.

Falk D (1993) Meningeal arterial patterns in great apes: implications for hominid vascular evolution. *Am J Phys Anthropol* 92, 81–97.

Falk D, Nicholls P (1992) Meningeal arteries in Rhesus macaques (*Macaca mulatta*): implications for vascular evolution in anthropoids. *Am J Phys Anthropol* 89, 299–308.

Ferrero EM (2011) *Anatomía comparada del sistema muscular de la extremidad posterior en primates superiores*. PhD Thesis, University of Valladolid, Spain.

Ferrero EM, Pastor JF, Fernandez F, Barbosa M, Diogo R, Wood B (2012) Comparative anatomy of the lower limb muscles of hominoids: attachments, relative weights, innervation and functional morphology. In: Hughes EF, Hill ME (eds.), Primates: Classification, evolution and behavior, p. 1–70. Hauppauge: Nova Science Publishers.

Fick R (1895a) Vergleichend-anatomische Studien an einem erwachsenen Orang-utang. *Arch Anat Physiol Anat Abt* 1895, 1–100.

Fick R (1895b) Beobachtungen an einem zweiten envachsenen Orang-Utang und einem Schimpansen. *Arch Anat Physiol Anat Abt* 1895, 289–318.

Forster A (1903) Die Insertion des Musculus semimembranosus. *Arch Anat Physiol Anat Abt* 1903, 257–320.

Frey H (1913) Der Musculus triceps surae in der Primatenreihe. *Morph Jahrb* 47, 1–192.

Fürbringer M (1875) *Beitrag zur Kenntnis* der Kehlkopfmuskulatur. Jena: Verlag von Hermann.

Gibbs S (1999) *Comparative soft tissue morphology of the extant Hominoidea, including Man*. Unpublished PhD Thesis, The University of Liverpool, Liverpool.

Gibbs S, Collard M, Wood BA (2000) Soft-tissue characters in higher primate phylogenetics. *Proc Natl Acad Sci US* 97, 11130–11132.

Gibbs S, Collard M, Wood BA (2002) Soft-tissue anatomy of the extant hominoids: a review and phylogenetic analysis. *J Anat* 200, 3–49.

Grönroos H (1903) Die musculi biceps brachii und latissimocondyloideus bei der affengattung *Hylobates* im vergleich mit den ensprechenden gebilden der anthropoiden und des menschen. *Abh Kön Preuss Akad Wiss Berlin* 1903, 1–102.

Groves CP (1986) Systematics of the great apes. In *Comparative Primate Biology: Systematics, Evolution and Anatomy, Vol. 1* (eds. Swindler DR, Erwin J), pp. 187–217. New York: A.R. Liss.

Groves CP (1995) *Revised character descriptions for Hominoidea*. Typescript, 9 pp.

Groves CP (2001) *Primate Taxonomy*. Washington, DC: Smithsonian Institution Press.

Hamada Y (1985) Primate hip and thigh muscles: comparative anatomy and dry weights. In *Primate Morphophysiology, Locomotor Analyses and Human Bipedalism* (ed. Kondo S), pp. 131–152. Tokyo: University of Tokyo Press.

Hanna JB, Schmitt D (2011) Comparative triceps surae morphology in primates: a review. *Anat Res Int* 2011, 191509.

Hartmann R (1886) *Anthropoid apes*. London: Keegan.

Hepburn D (1892) The comparative anatomy of the muscles and nerves of the superior and inferior extremities of the anthropoid apes: I-Myology of the superior extremity. *J Anat Physiol* 26, 149–186.

Hofër W (1892) Vergleichend-anatomische Studien uber die Nerven des Armes und der Hand bei den Affen und dem Menschen. *Munchener Med Abhandl* 30, 1–106.

Howell AB (1936a) Phylogeny of the distal musculature of the pectoral appendage. *J Morphol* 60, 287–315.

Howell AB (1936b) The phylogenetic arrangement of the muscular system. *Anat Rec* 66, 295–316.

Howell AB, Straus WL (1932) The brachial flexor muscles in primates. *Proc US Natl Mus* 80, 1–31.

Huber E (1930a) Evolution of facial musculature and cutaneous field of trigeminus—Part I. *Q Rev Biol* 5, 133–188.

Huber E (1930b) Evolution of facial musculature and cutaneous field of trigeminus—Part II. *Q Rev Biol* 5, 389–437.

Huber E (1931) *Evolution of facial musculature and expression.* Baltimore: The Johns Hopkins University Press.

Huntington GS (1903) Present problems of myological research and the significance and classification of muscular variations. *Am J Anat* 2, 157–175.

Huxley TH (1863) *Evidence as to Man's Place in Nature.* London: Williams and Norgate.

Huxley TH (1864) The structure and classification of the Mammalia. *Med Times Gazette* 1864, 398–468.

Huxley TH (1864) The structure and classification of the Mammalia. *Med Times Gazette* 1864, 398–468.

Huxley TH (1871) *The anatomy of vertebrated animals.* London; J. & A. Churchill.

Jordan J (1971a) Studies on the structure of the organ of voice and vocalization in the chimpanzee, Part 1. *Folia Morphol* 30, 99–117.

Jordan J (1971b) Studies on the structure of the organ of voice and vocalization in the chimpanzee, Part 2. *Folia Morphol* 30, 222–248.

Jordan J (1971c) Studies on the structure of the organ of voice and vocalization in the chimpanzee, Part 3. *Folia Morphol* 30, 323–340.

Jouffroy FK (1971) Musculature des membres. In *Traité de Zoologie, XVI: 3 (Mammifères)* (ed. Grassé PP), pp. 1–475. Paris: Masson et Cie.

Jouffroy FK, Lessertisseur J (1958) Notes sur la musculature de la main et du pied d'un orang-outan *(Pongo pygmaeus Hoppius). Bull Mus nat Hist Nat, Ser 2,* 30, 111–122.

Jouffroy FK, Lessertisseur J (1959) Reflexions sur les muscles contracteurs des doigts et des orteils (contrahentes digitorum) chez les primates. *Ann Sci Nat Zool, Ser 12,* 1, 211–235.

Jouffroy FK, Lessertisseur J (1960) Les spécialisations anatomiques de la main chez les singes à progression suspendue. *Mammalia* 24, 93–151.

Jouffroy FK, Saban R (1971) Musculature peaucière. In *Traité de Zoologie, XVI: 3 (Mammifères)* (ed. Grassé PP), pp. 477–611. Paris: Masson et Cie.

Kallner M (1956) Die muskulatur und die funktion des schultergurtels und der vorderextremitat des orang-utans. *Morph Jahrb* 97, 554–665.

Kaneff A (1959) Über die evolution des m. abductor pollicis longus und m. extensor pollicis brevis. *Mateil morphol Inst Bulg Akad Wiss* 3, 175–196.

Kaneff A (1968) Zur differenzierung des m. abductor pollicis biventer beim Menschen. *Gegen Morphol Jahrb* 112, 289–303.

Kaneff A (1969) Umbildung der dorsalen Daumenmuskeln beim Menschen. *Verh Anat Ges* 63, 625–636.

Kaneff A (1979) Évolution morphologique des musculi extensores digitorum et abductor pollicis longus chez l'Homme. I. Introduction, méthodologie, M. extensor digitorum. *Gegen Morphol Jahrb* 125, 818–873.

Kaneff A (1980a) Évolution morphologique des musculi extensores digitorum et abductor pollicis longus chez l'Homme. II. Évolution morphologique des m. extensor digiti minimi, abductor pollicis longus, extensor pollicis brevis et extensor pollicis longus chez l'homme. *Gegen Morphol Jahrb* 126, 594–630.

Kaneff A (1980b) Évolution morphologique des musculi extensores digitorum et abductor pollicis longus chez l'Homme. III. Évolution morphologique du m. extensor indicis chez l'homme, conclusion générale sur 1 'évolution morphologique des musculi extensores digitorum et abductor pollicis longus chez l'homme. *Gegen Morphol Jahrb* 126, 774–815.

Kaneff A (1986) Die Aufrichtung des Menschen und die mor-phologisches Evolution der Musculi extensores digitorum pedis unter dem Gesichtpunkt der evolutiven Myologie, Teil I. *Morph Jahrb* 132, 375–419.

Kaneff A, Cihak R (1970) Modifications in the musculus extensor digitorum lateralis in phylogenesis and in human ontogenesis. *Acta Anat Basel* 77, 583–604.

Kaplan EB (1958a) The iliotibial tract—clinical and morphological significance. *J Bone Jt Surg* 40A, 817–832.

Kaplan EB (1958b) Comparative anatomy of the extensor digitorum longus in relation to the knee joint. *Anat Rec* 131, 129–149.

Kawai K, Koizumi M, Honma S, Tokiyoshi A, Kodama K (2003) Derivation of the anterior belly of the digastric muscle receiving twigs from the mylohyoid and facial nerves. *Ann Anat* 185, 85–90.

Kawashima T, Yoshitomi S, Sasaki H (2007) Nerve fiber tracing of the branches to the coracobrachialis in an adult male orangutan. *Anat Histol Embryol* 36, 19–23.

Keith A (1894a) *The myology of the Catarrhini: a study in evolution.* Unpublished PhD thesis, University of Alberdeen, Alberdeen.

Keith A (1894b) Notes on a theory to account for the various arrangements of the flexor profundus digitorum in the hand and foot of primates. *J Anat Physiol* 28, 335–339.

Kleinschmidt A (1938) Die Schlund-und Kehlorgane des Gorillas "Bobby" unter besonderer—Berücksichtigung der gleichen Organe von Mensch und Orang—Ein Beitrag zur vergleichenden Anatomie des kehlkopfes. *Morphol Jahrb* 81, 78–157.

Kohlbrügge JHF (1896) Der larynx und die stimmbildung der Quadrumana. *Natuurk T Ned Ind* 55, 157–175.

Kohlbrügge JHF (1897) Muskeln und Periphere Nerven der Primaten, mit besonderer Berücksichtigung ihrer Anomalien. *Verh K Akad Wet Amsterdam Sec 2.* 5, 1–246.

Koizumi M, Sakai T (1995) The nerve supply to coracobrachialis in apes. *J Anat* 186, 395–403.

Körner O (1884) Beiträge zur vergleichenden Anatomie und Physiologie des Kehlkopfes der Säugethiere und des Menschen. *Abh Senckenb Naturforsch Ges* 13, 147–261.

Kumakura H (1989) Functional analysis of the biceps femoris muscle during locomotor behavior in some primates. *Am J Phys Anthropol* 79, 379–391.

Laitman JT (1977) *The Ontogenetic and phylogenetic development of the upper respiratory system and basicranium in man.* Unpublished PhD thesis, Yale University, New Haven.

Langer C (1879) Die Muskeln der Extremitäten des Orang als Grundlage einer vergleichend—myologischen Untersuchung. *Sitzugsber Math Nat Cl Kais Akad Wiss Wien* 79, 177–222.

Lartschneider J (1895): Zur vergleichende Anatomie des Diaphragma pelvis. *Sitzungsber Kaiserl Akad Wiss-Wien, Math Natunviss Kl* 104, 160–190.

Lessertisseur J (1958) Doit-on distinguer deux plans de muscles interosseux à la main et au pied des primates? *Ann Sci nat Zool Biol Anim* 20, 77–103.

Lewis OJ (1962) The comparative morphology of M. flexor accessorius and the associated long flexor tendons. *J Anat* 96, 321–333.

Lewis OJ (1964) The evolution of the long flexor muscles of the leg and foot. In *International Review of General and Experimental Zoology* (eds. Felts WJL, Harrison RJ), pp. 165–185. New York: Academic Press.

Lewis OJ (1965) The evolution of the Mm. interossei in the primate hand. *Anat Rec* 153, 275–287.

Lewis OJ (1966) The phylogeny of the cruropedal extensor musculature with special reference to the primates. *J Anat* 100, 865–880.

Lewis OJ (1989) Functional morphology of the evolving hand and foot. Oxford: Clarendon Press.

Lightoller GHS (1925) Facial muscles—the modiolus and muscles surrounding the rima oris with some remarks about the panniculus adiposus. *J Anat Physiol* 60, 1–85.

Lightoller GS (1928a) The facial muscles of three orang utans and two cercopithecidae. *J Anat* 63, 19–81.

Lightoller GS (1928b) The action of the m. mentalis in the expression of the emotion of distress. *J Anat* 62, 319–332.

Lightoller GS (1934) The facial musculature of some lesser primates and a *Tupaia*. *Proc Zool Soc Lond* 1934, 259–309.

Lightoller GS (1939) V. Probable homologues. A study of the comparative anatomy of the mandibular and hyoid arches and their musculature—Part I. Comparative myology. *Trans Zool Soc Lond* 24, 349–382.

Lightoller GS (1940a) The comparative morphology of the platysma: a comparative study of the sphincter colli profundus and the trachelo-platysma. *J Anat* 74, 390–396.

Lightoller GS (1940b) The comparative morphology of the m. caninus. *J Anat* 74, 397–402.

Loth E (1931) *Anthropologie des parties molles (muscles, intestins, vaisseaux, nerfs peripheriques)*. Paris: Mianowski-Masson et Cie.

Lunn HF (1948) The comparative anatomy of the inguinal ligament. *Anat Phys* 82, 58–67.

Lunn HF (1949) Observations on the mammalian inguinal region. *Proc Zool Soc* 118, 345–355.

Macalister A (1871) On some points in the myology of the chimpanzee and other primates. *Ann Mag Nat Hist* 7, 341–351.

Manners-Smith T (1908) A study of the cuboid and os peroneum in the primate foot. *J Anat Phys* 42, 397–414.

Mayer JC (1856) Zur Anatomie des Orang-utang und des Schimpansen. *Arch Naturgesch* 22, 279–304.

Michaëlis P (1903) Beitrage zur vergleichenden Myologie des *Cynocephalus babuin*, *Simia satyrus*, *Troglodytes niger*. *Arch Anat Phys Anat Abt* 1, 205–256.

Mijsberg WA (1923) Über den Bau des Urogenitalapparates bei den männlichen Primaten. *Verh K Akad Wet Amsterdam* 23, 1–92.

Miller RA (1932) Evolution of the pectoral girdle and forelimb in the primates. *Amer J Phys Anthropol* 17, 1–56.

Miller RA (1934) Comparative studies upon the morphology and distribution of the brachial plexus. *Am J Anat* 54, 143–175.

Miller RA (1945) The ischial callosities of primates. *Am J Anat* 76, 67–87.

Miller RA (1947) The inguinal canal of primates. *Am J Anat* 90, 117–142.

Miller RA (1952) The musculature of *Pan paniscus*. *Am J Anat* 91, 182–232.

Morimoto N, De León MSP, Nishimura T, Zollikofer CP (2011) Femoral Morphology and Femoropelvic Musculoskeletal Anatomy of Humans and Great Apes: A Comparative Virtopsy Study. *Anat Rec* 294, 1433–1445.

Myatt JP, Crompton RH, Thorpe SKS (2011) Hindlimb muscle architecture in non-human great apes and a comparison of methods for analysing inter-species variation. *J Anat* 219, 150–66.

Myatt JP, Schilling N, Thorpe SKS (2011) Distribution patterns of fibre types in the triceps surae muscle group of chimpanzees and orangutans. *J Anat* 218, 402–412.

Myatt JP, Crompton RH, Payne-Davis RC, Vereecke EE, Isler K, Savage R, D'Août K, Günther MM, Thorpe SK (2012) Functional adaptations in the forelimb muscles of non-human great apes. *J Anat* 220, 13–28.

Mysberg WA (1917) Über die Verbinderungen zwischen dem Sitzbeine und der Wirbelsäule bei den Säugetieren. *Anat Hefte* 54, 641–668.

Negus VE (1949) *The comparative anatomy and physiology of the larynx*. New York: Hafner Publishing Company.

Oishi M, Ogihara N, Endo H, Asari M (2008) Muscle architecture of the upper limb in the orangutan. *Primates* 49, 204–209.

Oishi M, Ogihara N, Endo H, Ichihara N, Asari M (2009) Dimensions of forelimb muscles in orangutans and chimpanzees. *J Anat* 215, 373–382.

Oishi M, Ogihara N, Endo H, Une Y, Ichihara N, Asari M, Amasaki H (2012) Muscle dimensions of the foot in the orangutan and the chimpanzee. *J Anat* 221, 311–317.

Owen R (1830-1831) On the anatomy of the orangutan (*Simia satyrus*, L.). *Proc Zool Soc Lond* 1, 4–5, 9–10, 28–29, 66–72.

Owen R (1868) *The Anatomy of Vertebrates, Vol. 3: Mammals*. London: Longmans, Green & Co.

Parsons FG (1898a) The muscles of mammals, with special relation to human myology, Lecture 1, The skin muscles and muscles of the head and neck. *J Anat Physiol* 32, 428–450.

Parsons FG (1898b) The muscles of mammals, with special relation to human myology: a course of lectures delivered at the Royal College of Surgeons of England—lecture II, the muscles of the shoulder and forelimb. *J Anat Physiol* 32, 721–752.

Payne RC (2001) *Musculoskeletal adaptations for climbing in hominoids and their role as exaptations for the acquisition of bipedalism*. Unpublished PhD thesis, The University of Liverpool, Liverpool.

Payne RC, Crompton RH, Isler K, Savage R, Vereecke EE, Gunther MM, Thorpe SKS, D'Aout K (2006) Morphological analysis of the hindlimb in apes and humans. I. Muscle architecture. J *Anat* 208, 709–724.

Pearson K, Davin AG (1921) On the sesamoids of the knee joint. *Biometrika* 13, 133–175, 350–400.

Plattner F (1923) Über die ventral-innervierte und die genuine rückenmuskulatur bei drei anthropomorphen (*Gorilla gina, Hylobates* und *Troglodytes niger*). *Morphol Jb* 52, 241–280.

Potau JM, Bardina X, Ciurana N, Camprubi D, Pastor JF, De Paz F, Barbosa M (2009) Quantitative analysis of the deltoid and rotator cuff muscles in humans and great Apes. *Int J Primatol* 30, 697–708.

Prejzner-Morawska A, Urbanowicz M (1971) The biceps femoris muscle in lemurs and monkeys. *Folia Morphol* 30, 9465–482.

Primrose A (1899) The anatomy of the orang-outang (*Simia satyrus*), an account of some of its external characteristics, and the myology of the extremities. *Trans Royal Can Inst* 6, 507–594.

Primrose A (1900) The anatomy of the orang-outang. *Univ Toronto Stud Anat Ser* 1, 1–94.

Ranke K (1897) Muskel-und Nervenvariationen der dorsalen elemente des Plexus ischiadicus der Primaten. *Arch Anthropol* 24, 117–144.

Rauwerdink GP (1993) Muscle fibre and tendon lengths in primate extremities. In *Hands of Primates* (eds. Preuschoft H, Chivers DJ), pp. 207–223. New York: Springer-Verlag.

Richmond BG (1993) Anatomy of the orang-utan, *Pongo pygmaeus abelii*. Unpublished work. State University of New York, Stony Brook.

Robinson JT, Freedman L, Sigmon BA (1972) Some aspects of pongid and hominid bipedality. *J Hum Evol* 1, 361–369.

Ruge G (1878a) Untersuchung uber die Extensorengruppe aus Unterschenkel und Füsse der Säugethiere. *Morphol Jahrb* 4, 592–643.

Ruge G (1878b) Zur vergleichenden Anatomie der tiefen Muskeln in der Fusssohle. *Morphol Jahrb* 4, 644–659.

Ruge G (1885) Über die Gesichtsmuskulatur der halbaffen. *Gegen Morph Jahrb* 11, 243–315.

Ruge G (1887a) *Untersuchungen uber die Gesichtsmuskeln der Primaten*. Leipzig: W. Engelmann.

Ruge G (1897) *Über das peripherische gebiet des nervus* facialis *boi wirbelthieren*. Leipzig: Festschr f Gegenbaur.

Saban R (1968) Musculature de la tête. In Traité de Zoologie, XVI: 3 (Mammifères) (ed. Grassé PP), pp. 229–472. Paris: Masson et Cie.

Sakka M (1973) Anatomie comparée de l'écaille de l'occipital (squama occipitalis P.N.A.) et des muscles de la nuque chez l'homme et les pongidés, II Partie, Myologie. *Mammalia* 37, 126–180.

Sandifort G (1840) Ontleedkundige beschouwing van een' volwassen Orang-oetan (*Simia satyrus*, Linn.) van het mannelijk geslacht. Verhandelingen over de Natuurlijke geschiedenis der

Nederlandische overzeesche bezittingen, door de leden der Natuurkundige commissie in Indië en andere. *Schrijvers (Zool)* 3, 29–56.

Sarmiento EE (1983) The significance of the heel process in anthropoids. *Int J Primatol* 4, 127–152.

Schneider R (1964) Der Larynx der Säugetiere. *Handbuch der Zoologie* 5, 1–128.

Schück AC (1913a) Beiträge zur Myologie der Primaten, I—der m. lat. dorsi und der m. latissimo-tricipitalis. *Morphol Jahrb* 45, 267–294.

Schück AC (1913b) Beiträge zur Myologie der Primaten, II—1 die gruppe sterno-cleido-mastoideus, trapezius, omo-cervicalis, 2 die gruppe levator scapulae, rhomboides, serratus anticus. *Morphol Jahrb* 46, 355–418.

Seiler R (1971a) A comparison between the facial muscles of Catarrhini with long and short muzzles. *Proc 3rd Int Congr Primat Zürich 1970, vol l, Basel: Karger*, 157–162.

Seiler R (1971b) Facial musculature and its influence on the facial bones of catarrhine Primates, I. *Morphol Jahrb* 116, 122–142.

Seiler R (1971c) Facial musculature and its influence on the facial bones of catarrhine Primates, II. *Morphol Jahrb* 116, 147–185.

Seiler R (1971d) Facial musculature and its influence on the facial bones of catarrhine Primates, III. *Morphol Jahrb* 116, 347–376.

Seiler R (1971e) Facial musculature and its influence on the facial bones of catarrhine Primates, IV. *Morphol Jahrb* 116, 456–481.

Seiler R (1973) On the function of facial muscles in different behavioral situations—a study based on muscle morphology and electromyography. *Am J Phys Anthropol* 38, 567–71.

Seiler R (1974a) Muscles of the external ear and their function in man, chimpanzees and *Macaca*. *Morphol Jahrb* 120, 78–122.

Seiler R (1974b) Particularities in facial muscles of *Daubentonia madagascariensis* (Gmelin 1788). *Folia Primatol* 22, 81–96.

Seiler R (1975) On the facial muscles in *Perodicticus potto* and *Nycticebus coucang*. *Folia Primatol* 23, 275–289.

Seiler R (1976) Die Gesichtsmuskeln. In *Primatologia, Handbuch der Primatenkunde, Bd. 4, Lieferung 6* (eds. Hofer H, Schultz AH, Starck D), pp. 1–252. Basel: Karger.

Seiler R (1977) Morphological and functional differentiation of muscles—studies on the m. frontalis, auricularis superior and auricularis anterior of primates including man. *Verh Anat Ges* 71, 1385–1388.

Seiler R (1979a) Criteria of the homology and phylogeny of facial muscles in primates including man, I, Prosimia and Platyrrhina. *Morphol Jahrb* 125, 191–217.

Seiler R (1979b) Criteria of the homology and phylogeny of facial muscles in primates including man, II, Catarrhina. *Morphol Jahrb* 125, 298–323.

Seiler R (1980) Ontogenesis of facial muscles in primates. *Morphol Jahrb* 126, 841–864.

Shoshani J (1986) *On the Phylogenetic Relationships among Paenungulata and within Elephantidae as Demonstrated by Molecular and Osteological Evidence.* Unpublished PhD thesis, Wayne State University, Detroid.

Shoshani J, Groves CP, Simons EL, Gunnell GF (1996) Primate phylogeny: morphological vs molecular results. *Mol Phylogenet Evol* 5, 102–154.

Shrewsbury MM, Marzke MM, Linscheid RL, Reece SP (2003) Comparative morphology of the pollical distal phalanx. *Am J Phys Anthropol* 121, 30–47.

Shrivastava RK (1978) *Anatomie comparée du muscle deltoïde et son innervation dans la série des mammifères.* Unpublished Phd thesis, Université de Paris, Paris.

Sigmon BA (1974) A functional analysis of pongid hip and thigh musculature. *J Hum Evol* 3, 161–185.

Smith WC (1923) The levator ani muscle; its structure in man, and its comparative relationships. *Anat Rec* 26, 175–204.

Sneath RS (1955) The insertion of the biceps femoris. *J Anat* 89, 550–553.

Sonntag CF (1924a) On the anatomy, physiology, and pathology of the orang-outan. *Proc Zool Soc Lond* 24, 349–450.

Sonntag CF (1924b) *The morphology and evolution of the apes and man.* London: John Bale Sons and Danielsson, Ltd.

Starck D, Schneider R (1960) Respirationsorgane. In *Primatologia III/2* (eds. Hofer H, Schultz AH, Starck D), pp. 423–587. Basel: Karger.

Stern JT (1972) Anatomical and functional specializations of the human gluteus maximus. *Anta J Phys Anthropol* 36, 315–340.

Stewart TD (1936) The musculature of the anthropoids, I, neck and trunk. *Am J Phys Anthropol* 21, 141–204.

Straus WL (1941a) The phylogeny of the human forearm extensors. *Hum Biol* 13, 23–50.

Straus WL (1941b) The phylogeny of the human forearm extensors (concluded). *Hum Biol* 13, 203–238.

Straus WL (1942a) The homologies of the forearm flexors: urodeles, lizards, mammals. *Am J Anat* 70, 281–316.

Straus WL (1942b) Rudimentary digits in primates. *Q Rev Biol* 17, 228–243.

Sullivan WE, Osgood CW (1925) The facialis musculature of the orang, *Simia satyrus. Anat Rec* 29, 195–343.

Sullivan WE, Osgood CW (1927) The musculature of the superior extremity of the orang-utan, *Simia satyrus. Anat Rec* 35, 193–239.

Susman RL, Nyati L, Jassal MS (1999) Observations on the pollical palmar interosseus muscle (of Henle). *Anat Rec* 254, 159–165.

Testut L (1883) Le long fléchisseur propre du pouce chez l'homme et les singes. *Bull Soc Zool Fr* 8, 164–185.

Testut L (1884) *Les anomalies musculaires chez l'homme expliquées par l'anatomie comparée et leur importance en anthropologie.* Paris: Masson.

Thompson P (1901) On the arrangement of the fasciae of the pelvis and their relationship to the levator ani. *J Anat Phys* 35, 127–141.

Thomson A (1915) On the presence of genial tubercles on the mandible of man, and their suggested association with the faculty of speech. *J Anat Physiol* 50, 43–74.

Tocheri MW, Orr CM, Jacofsky MC, Marzke MW (2008) The evolutionary history of the hominin hand since the last common ancestor of Pan and Homo. *J Anat* 212, 544–562.

Toldt C (1905) Der Winkelforsatz des Unterkiefers beim Menschen und bei den Saugeticren und die Beziehungen der Kaumuskeln zu demselben II Teil. *Sitz Kaiserl Akad Wiss Wien Mathem-naturw Klasse Bd* 114, 315–476.

Traill TS (1821) Observations on the anatomy of the orangutan. *Mem Weiner Nat Hist Soc Edinburgh* 3, 1–49.

Tschachmachtschjan H (1912) Über die Pectoral- und Abdominal- musculatur und über die Scalenus-Gruppe bei Primataten. *Morph Jb* 44, 297–370.

Tuttle RH (1967) Knuckle-walking and the evolution of hominoid hands. *Am J Phys Anthrop* 26, 171–206.

Tuttle RH (1969) Quantitative and functional studies on the hands of the Anthropoidea, I, the Hominoidea. *J Morphol* 128, 309–363.

Tuttle RH (1970) Postural, propulsive and prehensile capabilities in the cheiridia of chimpanzees and other great apes. In *The Chimpanzee, Vol. 2* (ed. Bourne GH), pp. 167–263. Basel: Karger.

Tuttle RH (1972a) Relative mass of cheiridial muscles in catarrhine primates. In *The Functional and Evolutionary Biology of Primates* (ed. Tuttle RH), pp. 262–291. Chicago: Aldine-Atherdon.

Tuttle RH, Basmajian JV (1976) Electromyography of pongid shoulder muscles and hominoid evolution I—retractors of the humerus and rotators of the scapula. *Yearbook Phys Anthropol* 20, 491–497.

Tuttle RH, Basmajian JV (1978a) Electromyography of pongid shoulder muscles II—deltoid, rhomboid and "rotator cuff". *Am J Phys Anthropol* 49, 47–56.

Tuttle RH, Basmajian JV (1978b) Electromyography of pongid shoulder muscles III—quadrupedal positional behavior. *Am J Phys Anthropol* 49, 57–70.

Tuttle RH, Cortright G (1988) Positional behavior, adaptive complexes, and evolution. In Orang-Utan Biology (ed. Schwartz JH), pp. 311–330. Oxford: Oxford University Press.

Tuttle RH, Velte MJ, Basmajian JV (1983) Electromyography of brachial muscles in *Pan troglodytes* and *Pongo pygmaeus*. *Am J Phys Anthropol* 61, 75–83.

Tuttle RH, Hollowed JR, Basmajian JV (1992) Electromyography of pronators and supinators in great apes. *Am J Phys Anthropol* 87, 215–26.

Vallois H (1914) *Étude anatomique de l'articulation du genou ches les Primates*. Montpellier: L'Abeille.

Van den Broek AJ (1914) Studien zur Morphologie des Primatenbeckens. *Morphol Jahrb* 49, 1–118.

Van Westrienen A (1907) Das Kniegelenk der Primaten, mit besonderer Berücksichtigung der Anthropoiden. *Petrus Camper* 4, 1–60.

Vrolik W (1841) *Recherches d' anatornie comparé, sur le chimpanzé*. Amsterdam: Johannes Miller.

Wall CE, Larson SG, Stern JT (1994) EMG of the digastric muscle in gibbon and orangutan: functional consequences of the loss of the anterior digastric in orangutans. *Am J Phys Anthropol* 94, 549–567.

Whitehead PF (1993) Aspects of the anthropoid wrist and hand. In *Postcranial Adaptation in Nonhuman Primates* (ed. Gebo DL), pp 96–120. DeKalb: Northern Illinois University Press.

Wilkinson JL (1953) The insertions of the flexores pollicis longus et digitorum profundus. *J Anat* 87, 75–88.

Wind J (1970) *On the phylogeny and ontogeny of the human larynx*. Groningen: Wolters-Noordhoff.

Winkler LA (1989) Morphology and relationships of the orangutan fatty cheek pads. *Am J Primatol* 17, 305–320.

Winkler LA (1991) Morphology and variability of masticatory structures in the orangutan. *Int J Primat* 12, 45–65.

Wood Jones F (1920) *The principles of anatomy as seen in the hand*. London: J & A Churchill.

Yirga S (1987) Interrelation between ischium, thigh extending muscles and locomotion in some primates. *Primates* 28, 79–86.

Yoshikawa T (1961) The lamination of the m. masseter of the crab-eating monkey, orang-utan and gorilla. *Primates* 3, 81.

Zihlman AL, Brunker L (1979) Hominid bipedalism: then and now. *Yearb Phys Anthropol* 22, 132–162.

Zihlman AL, Mcfarland RK, Underwood CE (2011) Functional Anatomy and Adaptation of Male Gorillas (*Gorilla gorilla gorilla*) With Comparison to Male Orangutans (*Pongo pygmaeus*). *Anat Rec* 294, 1842–1855.

Appendix II
Literature Cited, not Including Information about the Muscles of Orangutans

Diogo R (2004a) *Morphological evolution, aptations, homoplasies, constraints, and evolutionary trends: catfishes as a case study on general phylogeny and macroevolution*. Enfield: Science Publishers.

Diogo R (2004b) Muscles versus bones: catfishes as a case study for an analysis on the contribution of myological and osteological structures in phylogenetic reconstructions. *Anim Biol* 54, 373–391.

Diogo R (2007) *On the origin and evolution of higher-clades: osteology, myology, phylogeny and macroevolution of bony fishes and the rise of tetrapods*. Enfield: Science Publishers.

Diogo R (2008) Comparative anatomy, homologies and evolution of the mandibular, hyoid and hypobranchial muscles of bony fish and tetrapods: a new insight. *Anim Biol* 58, 123–172.

Diogo R (2009) The head musculature of the Philippine colugo (Dermoptera: *Cynocephalus volans*), with a comparison to tree-shrews, primates and other mammals. *J Morphol* 270, 14–51.

Diogo R, Abdala V (2007) Comparative anatomy, homologies and evolution of the pectoral muscles of bony fish and tetrapods: a new insight. *J Morphol* 268, 504–517.

Diogo R, Potau JM, Pastor JF, de Paz FJ, Ferrero EM, Bello G, Barbosa M, Wood B (2010) Photographic and Descriptive Musculoskeletal Atlas of *Gorilla*—with notes on the attachments, variations, innervation, synonymy and weight of the muscles. Oxford: Taylor & Francis.

Diogo R, Potau JM, Pastor JF, de Paz FJ, Ferrero EM, Bello G, Barbosa M, Aziz MA, Burrows AM, Arias-Martorell J, Wood B (2012) Photographic and Descriptive Musculoskeletal Atlas of Gibbons and Siamangs (*Hylobates*) - with notes on the attachments, variations, innervation, synonymy and weight of the muscles. Oxford: Taylor & Francis.

Diogo R, Potau JM, Pastor JF, de Paz FJ, Ferrero EM, Bello G, Barbosa M, Aziz MA, Burrows AM, Arias-Martorell J, Wood B (2013) Photographic and Descriptive Musculoskeletal Atlas of Chimpanzees (*Pan*)—with notes on the attachments, variations, innervation, synonymy and weight of the muscles. Oxford: Taylor & Francis.

Moore KL, Dalley AF (1999) *Clinically Oriented Anatomy, 4th ed*. Philadelphia: Lippincott, Williams & Wilkins.

Mustafa MA (2006) Neuroanatomy. *10th National Congress of Anatomy Bordum, Turkey, September 5*, 6–10.

Netter FH (2006) *Atlas of human anatomy (4th ed.)*. Philadelphia: Saunders.

Terminologia Anatomica (1998) *Federative Committee on Anatomical Terminology*. Stuttgart: Georg Thieme. New York: Columbia University Press.

Wood J (1870) On a group of varieties of the muscles of the human neck, shoulder, and chest, with their transitional forms and homologies in the Mammalia. *Philos Trans R Soc Lond* 160, 83–116.

Index

Deltoideus 31
Depressor anguli oris 19
Depressor glabellae 16
Depressor helicis 11
Depressor labii inferioris 19
Depressor septi nasi 17, 18
Depressor supercilii 15
Depressor tarsi 11
Detrusor vesicae 60
Diaphragma 57
Digastricus anterior 5
Digastricus posterior 7
Dorsoepitrochlearis 33

E

Epitrochleoanconeus 38
Erector spinae 54, 55
Extensor brevis digitorum manus 49
Extensor carpi radialis brevis 46
Extensor carpi radialis longus 46
Extensor carpi ulnaris 47
Extensor communis pollicis et indicis 50
Extensor digiti III proprius 49
Extensor digiti minimi 48
Extensor digiti quarti 48
Extensor digitorum 48
Extensor digitorum brevis 74
Extensor digitorum longus 70
Extensor hallucis brevis 75
Extensor hallucis longus 71
Extensor indicis 49
Extensor pollicis brevis 50
Extensor pollicis longus 50

F

Fibularis brevis 71
Fibularis longus 72
Fibularis tertius 71
Flexor brevis profundus 1 42, 43
Flexor brevis profundus 2 41, 42
Flexor brevis profundus 10 42
Flexor carpi radialis 39
Flexor carpi ulnaris 38
Flexor caudae 59
Flexor digiti minimi brevis 42, 45, 78
Flexor digitorum brevis 76
Flexor digitorum brevis manus 40
Flexor digitorum longus 73
Flexor digitorum profundus 36
Flexor digitorum superficialis 37

Flexor hallucis brevis 78
Flexor hallucis longus 73
Flexor pollicis brevis 42
Flexor pollicis brevis and flexor brevis
    profundus 2 43
Flexor pollicis longus 37
Flexores breves profundi 43
Frontalis 12

G

Gastrocnemius 72
Gemellus inferior 63
Gemellus superior 63
Genio-epiglotticus 25
Genioglossus 24
Genio-hyo-epiglotticus 25
Genio-hyoglossus 24
Geniohyoideus 24
Glosso-epiglotticus 25
Gluteus maximus 61
Gluteus medius 62
Gluteus minimus 62
Gracilis 68

H

Helicis 11
Hyo-epiglotticus 25
Hyoglossus 25

I

Iliacus 61
Iliocapsularis 69
Iliococcygeus 59
Iliocostalis 54
Incisivus labii inferioris 19
Incisivus labii superior 19
Incisurae Santorini 11
Incisurae terminalis 11
Infraorbitalis 15
Infraspinatus 31
Intercapitulares 41
Intercartilagineus 11
Intercostales externi 54
Intercostales interni 54
Interdigitales 43
Intermandibularis anterior 5
Intermetacarpales 43
Interossei accessorii 43
Interossei dorsales 43, 79

About The Authors

Rui Diogo is an Assistant Professor at the Howard University College of Medicine and a Resource Faculty at the Center for the Advanced Study of Hominid Paleobiology of George Washington University (US). He is the author or co-author of numerous publications, and the co-editor of the books *Catfishes* and *Gonorynchiformes and ostariophysan interrelationships—a comprehensive review*. He is the sole author or first author of several monographs, including the three atlases of gorillas, chimpanzees and hylobatids and the books *Morphological evolution, aptations, homoplasies, constraints and evolutionary trends—catfishes as a case study on general phylogeny and macroevolution*, *The origin of higher clades—osteology, myology, phylogeny and evolution of bony fishes and the rise of tetrapods*, *Muscles of vertebrates—comparative anatomy, evolution, homologies and development*, and *Comparative anatomy and phylogeny of primate muscles and human evolution*.

Josep Potau is Professor at the Department of Anatomy and Embryology of the University of Barcelona (Spain) and is the director of the University's Center for the Study of Comparative and Evolutionary Anatomy. His current research focus on the analysis of functional and anatomical adaptations associated with the evolution of different types of locomotion and of the upper limb musculature within primates. He has published several papers and book chapters on functional and comparative anatomy.

Juan Pastor is Professor at the Department of Anatomy of the University of Valladolid (Spain) and is the director of the University's Anatomical Museum, which houses the largest comparative osteological collection in Spain. He published several papers on comparative anatomy and anthropology.

Félix de Paz is Professor at the Department of Anatomy of the University of Valladolid and is a member of the Royal Academy of Medicine and Surgery of Valladolid (Spain). He has published several papers on comparative anatomy and anthropology.

Mercedes Barbosa is Professor at the Department of Anatomy of the University of Valladolid (Spain), and is a member of the Anatomical Society of Spain. She published several papers on physical anthropology.

Eva Ferrero is a biologist who graduated from the University of León (Spain) and obtained her PhD at the University of Valladolid (Spain) where she is undertaking research on the comparative anatomy of primates and other mammals.

Gaëlle Bello is a biologist who graduated from the University of La Coruña (Spain). She undertook a Master in Primatology at the University of Barcelona (Spain), and is now undertaking a PhD at the University of Barcelona that focuses on the evolution of the scapula within Primates and its adaptations to different types of locomotion.

M. Ashraf Aziz is Professor of Anatomy at the Department of Anatomy of Howard University College of Medicine (USA). His research focuses on the comparative gross and developmental morphology of modern human aneuploidy syndromes, the evolution of the muscles supplied by the trigeminal nerve and of the arm and hand muscles in non-human and human primates and the value of human cadaver dissections/prosections in the age of digital information systems. He has published numerous papers in international journals, including *Teratology*, *American Journal of Physical Anthropology*, *Journal of Anatomy*, *The Anatomical Record* and *Primates*.

Julia Arias-Martorell is a biologist now undertaking a PhD at the University of Barcelona (Spain) that focuses on functional morphology and variability of the forelimbs of the hominoids related to the diverse locomotor repertoires of the members of this clade, and on the enhancement of tridimensional techniques to study evolution.

Bernard Wood is University Professor of Human Origins and Director of the Center for the Advanced Study of Hominid Paleobiology at George Washington University (USA). His edited publications include *Food Acquisition and Processing in Primates* and *Major Topics in Primate and Human Evolution* and he is the author of *The Evolution of Early Man, Human Evolution, Koobi Fora Research Project—Hominid Cranial Remains (Vol. 4), Human Evolution—A Very Short Introduction*. He is the editor of the *Wiley-Blackwell Encyclopedia of Human Evolution*.

Color Plate Section

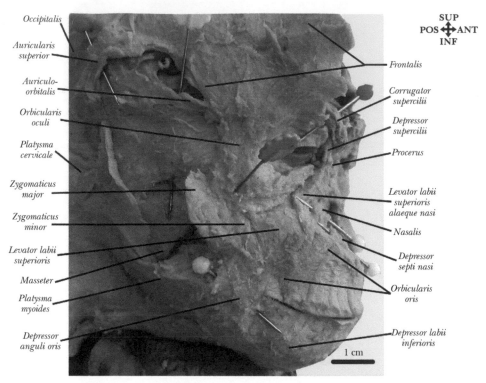

Fig. 1 *Pongo pygmaeus* (HU PP1, neonate male): lateral view of the right facial muscles. In this figure and the remaining figures of the atlas, the names of the muscles are in italics, and SUP, INF, ANT, POS, MED, LAT, VEN, DOR, PRO and DIS refer to superior, inferior, anterior, posterior, medial, lateral, ventral, dorsal, proximal and distal, respectively (in the sense the terms are applied to pronograde tetrapods: see Methodology and Material).

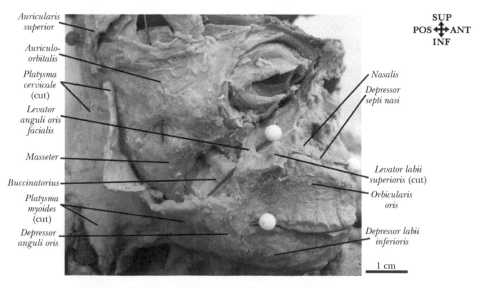

Fig. 2 *Pongo pygmaeus* (HU PP1, neonate male): lateral view of the deep left facial muscles; the levator labii superioris is reflected in order to show the levator anguli oris facialis.

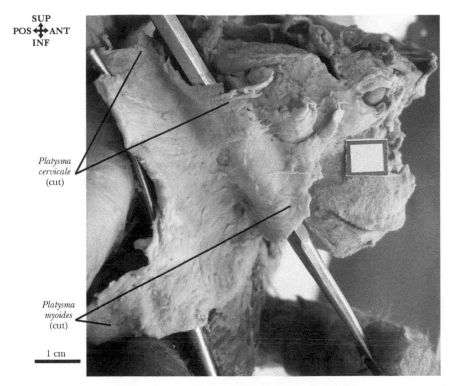

SUP
POS ⬌ ANT
INF

Platysma cervicale (cut)

Platysma myoides (cut)

1 cm

Fig. 3 *Pongo pygmaeus* (HU PP1, neonate male): lateral view of the right platysma myoides and platysma cervicale.

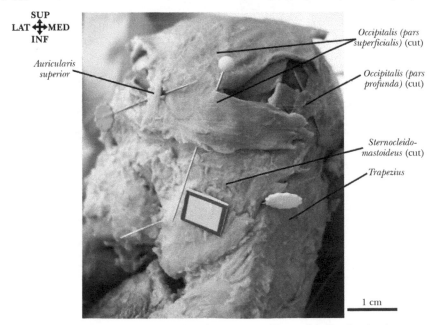

SUP
LAT ⬌ MED
INF

Auricularis superior

Occipitalis (pars superficialis) (cut)

Occipitalis (pars profunda) (cut)

Sternocleido-mastoideus (cut)

Trapezius

1 cm

Fig. 4 *Pongo pygmaeus* (HU PP1, neonate male): dorsolateral view of the occipitalis, showing its pars profunda and pars superficialis.

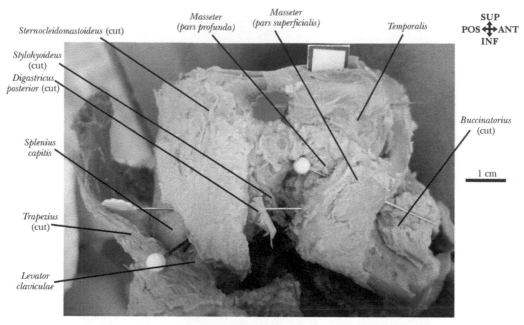

Fig. 5 *Pongo pygmaeus* (HU PP1, neonate male): lateral view of the left facial muscles; the levator labii superioris is reflected in order to show the levator anguli oris facialis.

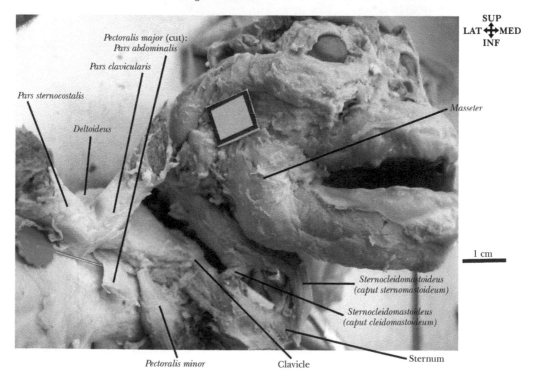

Fig. 6 *Pongo pygmaeus* (HU PP1, neonate male): ventrolateral view of the sternocleidomastoideus.

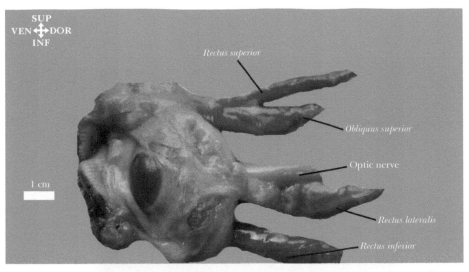

Fig. 7 *Pongo pygmaeus* (VU PP1, adult female): lateral view of the left eye ball and extraocular muscles.

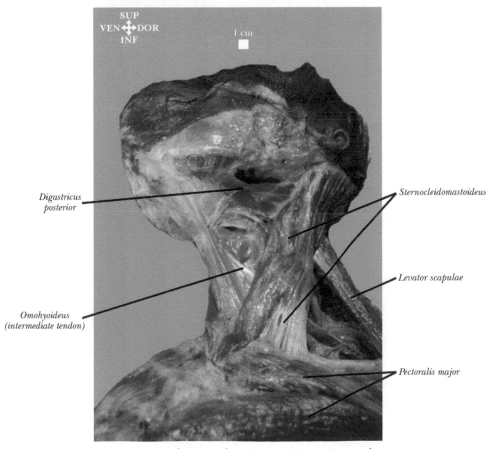

Fig. 8 *Pongo pygmaeus* (VU PP1, adult female): ventrolateral view of the neck musculature.

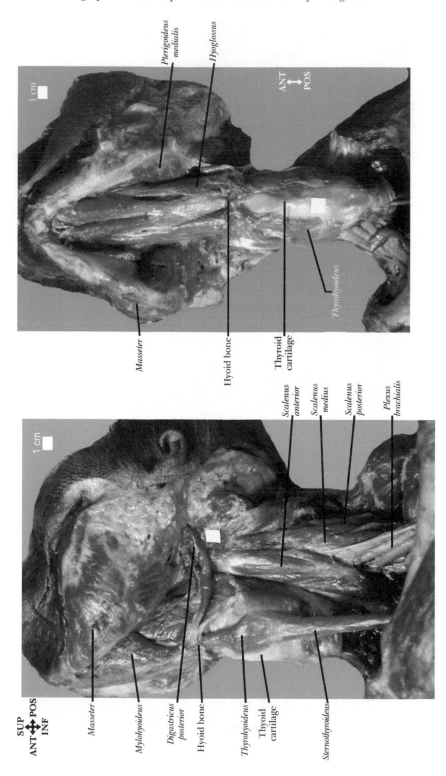

Fig. 10 *Pongo pygmaeus* (VU PP1, adult female): ventral view of the deep muscles of the neck.

Fig. 9 *Pongo pygmaeus* (VU PP1, adult female): ventrolateral view of the neck musculature after removal of the sternocleidomastoideus.

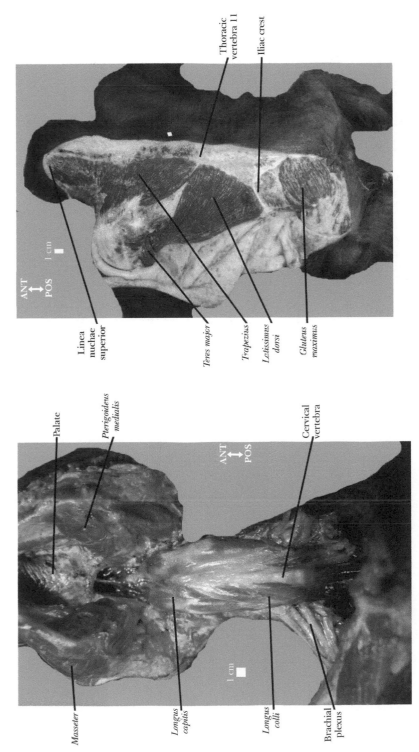

Fig. 12 *Pongo pygmaeus* (VU PP1, adult female): dorsal view of the superficial back musculature.

Fig. 11 *Pongo pygmaeus* (VU PP1, adult female): ventral view of the prevertebral musculature.

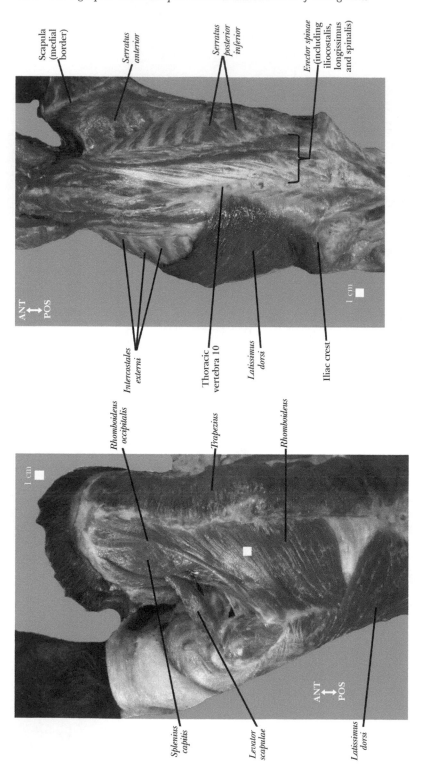

Fig. 13 *Pongo pygmaeus* (VU PP1, adult female): dorsal view of the back musculature after removal of the trapezius.

Fig. 14 *Pongo pygmaeus* (VU PP1, adult female): dorsal view of the deep musculature of the back.

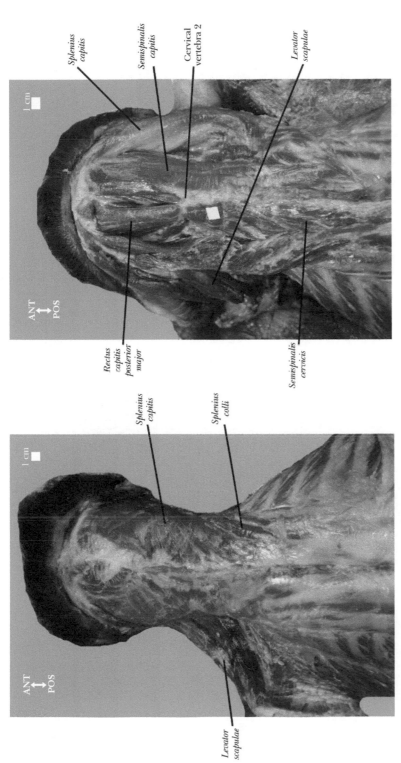

Fig. 16 *Pongo pygmaeus* (VU PP1, adult female): dorsal view of the deep back musculature.

Fig. 15 *Pongo pygmaeus* (VU PP1, adult female): dorsal view of the deep back musculature.

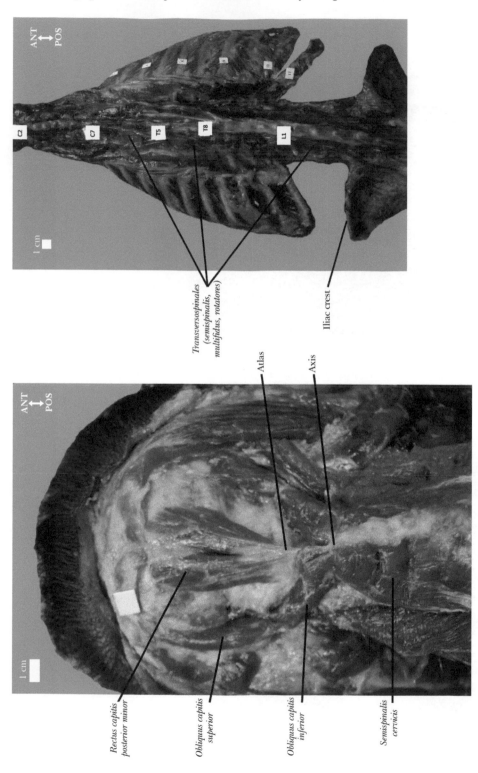

Fig. 18 *Pongo pygmaeus* (VU PP1, adult female): dorsal view of the deepest back muscles; C2, C7, T5, T8 and L1 refer to the cervical vertebrae 2 and 7, thoracic vertebrae 5 and 8, and lumbar vertebra 1, respectively.

Fig. 17 *Pongo pygmaeus* (VU PP1, adult female): dorsal view of the deepest back musculature.

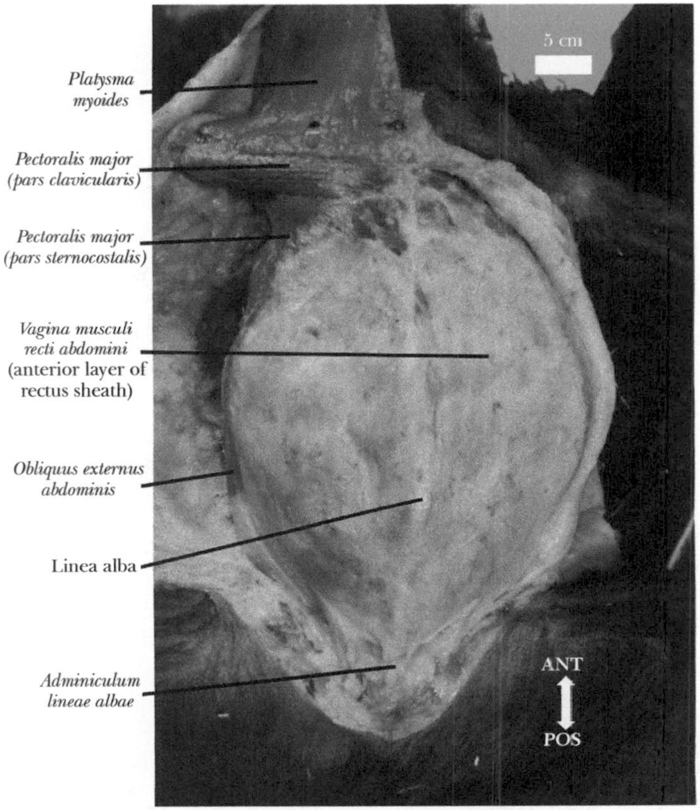

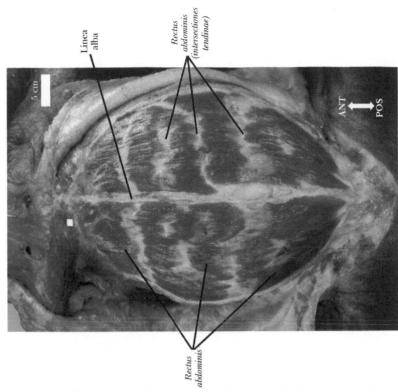

Fig. 20 *Pongo pygmaeus* (VU PP1, adult female): ventral view of the abdominal musculature after removal of the anterior layer of the rectus sheath.

Linea alba

Rectus abdominis (intersectiones tendinae)

Rectus abdominis

ANT POS

5 cm

Fig. 19 *Pongo pygmaeus* (VU PP1, adult female): ventral view of the abdominal musculature.

Platysma myoides

Pectoralis major (pars clavicularis)

Pectoralis major (pars sternocostalis)

Vagina musculi recti abdomini (anterior layer of rectus sheath)

Obliquus externus abdominis

Linea alba

Adminiculum lineae albae

ANT POS

5 cm

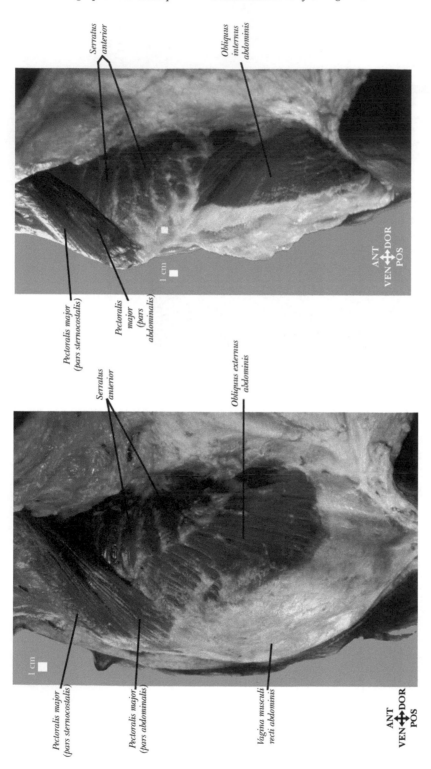

Fig. 22 *Pongo pygmaeus* (VU PP1, adult female): lateral view of the left abdominal musculature after removal of the obliquus externus abdominis and of the rectus abdominis.

Fig. 21 *Pongo pygmaeus* (VU PP1, adult female): lateral view of the superficial left abdominal musculature.

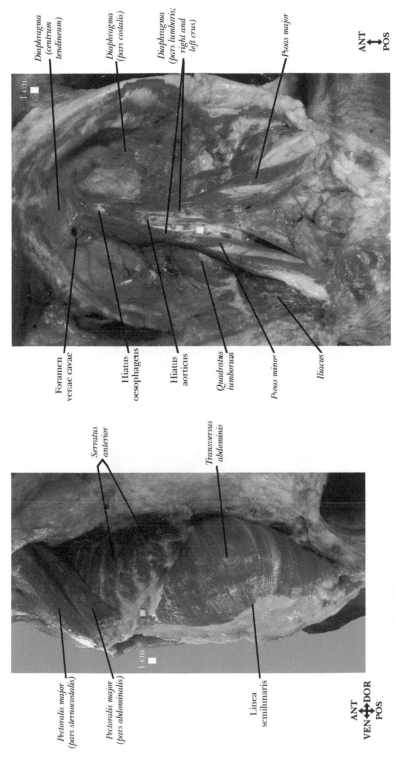

Fig. 24 *Pongo pygmaeus* (VU PP1, adult female): diaphragma and ventral view of the posterior abdominal musculature after removal of the viscerae.

Fig. 23 *Pongo pygmaeus* (VU PP1, adult female): lateral view of the left abdominal musculature after removal of the rectus abdominis, obliquus externus abdominis and obliquus internus abdominis.

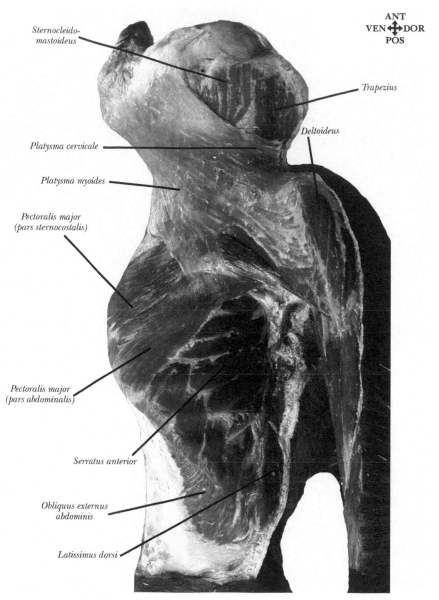

Fig. 25 *Pongo pygmaeus* (VU PP1, adult female): ventrolateral view of the head, neck, shoulder and trunk muscles.

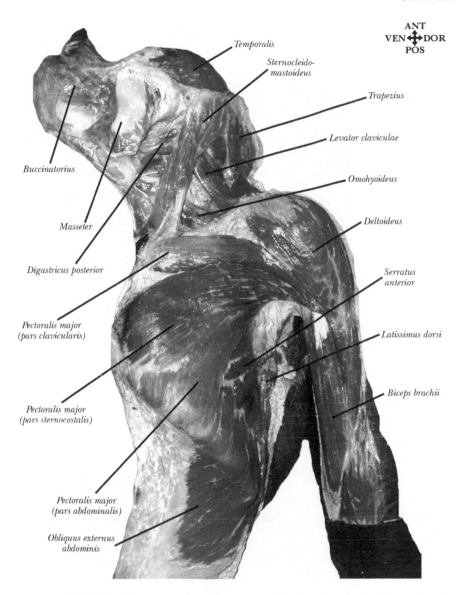

Fig. 26 *Pongo pygmaeus* (VU PP1, adult female): ventrolateral view of the head, neck, shoulder and trunk muscles after removal of platysma myoides and platysma cervicale.

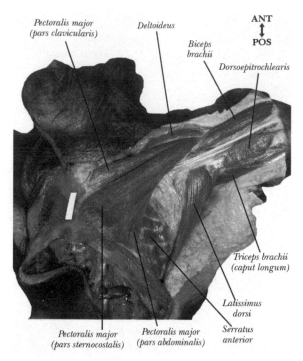

Fig. 27 *Pongo pygmaeus* (VU PP2, adult female): ventrolateral view of the shoulder and trunk muscles.

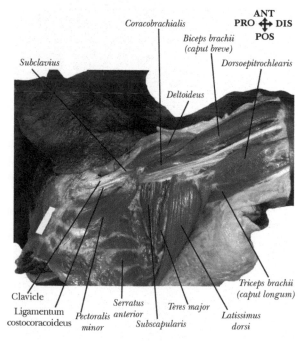

Fig. 28 *Pongo pygmaeus* (VU PP2, adult female): ventrolateral view of the shoulder and trunk muscles after removal of pectoralis major.

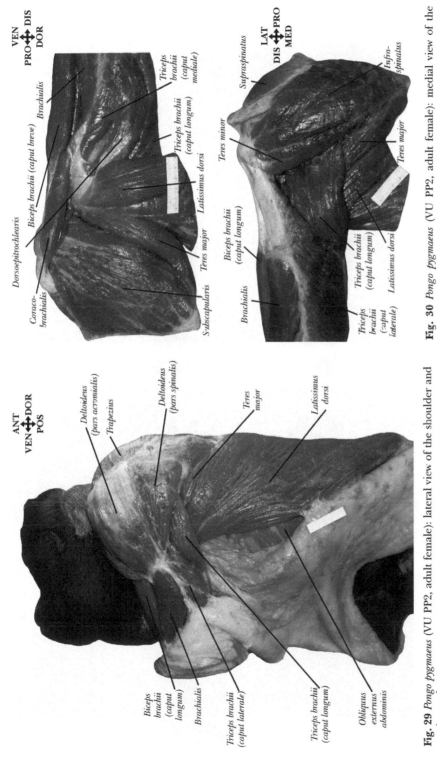

Fig. 29 *Pongo pygmaeus* (VU PP2, adult female): lateral view of the shoulder and trunk muscles.

Fig. 30 *Pongo pygmaeus* (VU PP2, adult female): medial view of the shoulder and arm muscles (on top); dorsolateral view of these muscles after removal of deltoideus (on bottom).

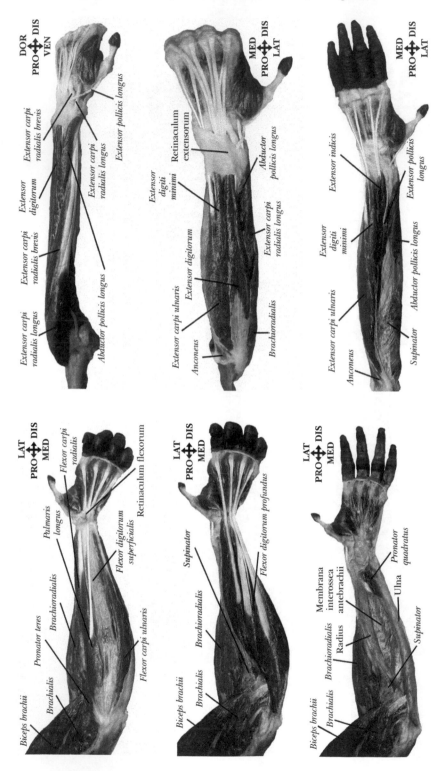

Fig. 31 *Pongo pygmaeus* (VU PP2, adult female): ventral view of superficial (on top), deeper (on center) and deepest (on bottom) forearm muscles.

Fig. 32 *Pongo pygmaeus* (VU PP2, adult female): lateral view of the forearm muscles after removal of brachioradialis (on top) and dorsal view of the superficial (on center) and deeper (on bottom) forearm muscles.

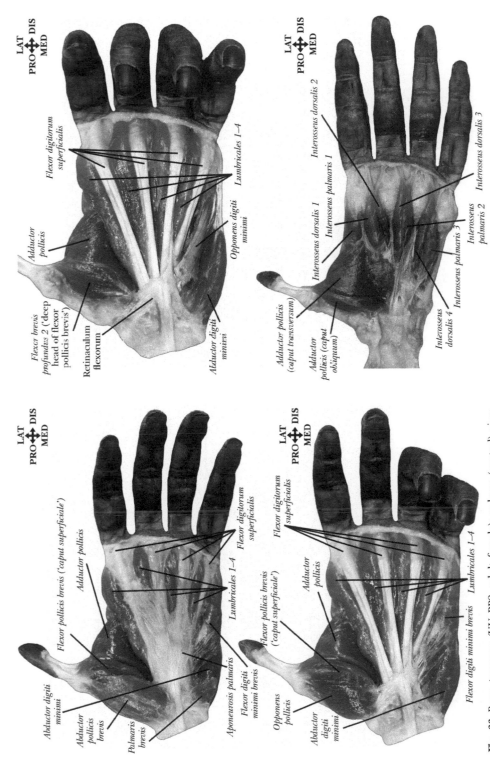

Fig. 34 *Pongo pygmaeus* (VU PP2, adult female): palmar (ventral) view of the deep (on top) and deeper (on bottom) hand muscles.

Fig. 33 *Pongo pygmaeus* (VU PP2, adult female): palmar (ventral) view of the hand muscles before (on top) and after (on bottom) removal of abductor pollicis brevis and palmaris brevis.

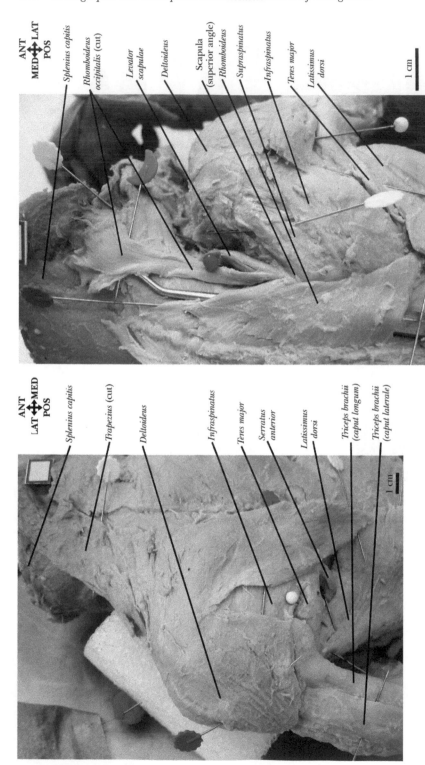

Fig. 35 *Pongo pygmaeus* (HU PP1, neonate male): dorsal view of the left pectoral and arm musculature.

Fig. 36 *Pongo pygmaeus* (HU PP1, neonate male): dorsolateral view of the right pectoral and arm musculature after removing the trapezius.

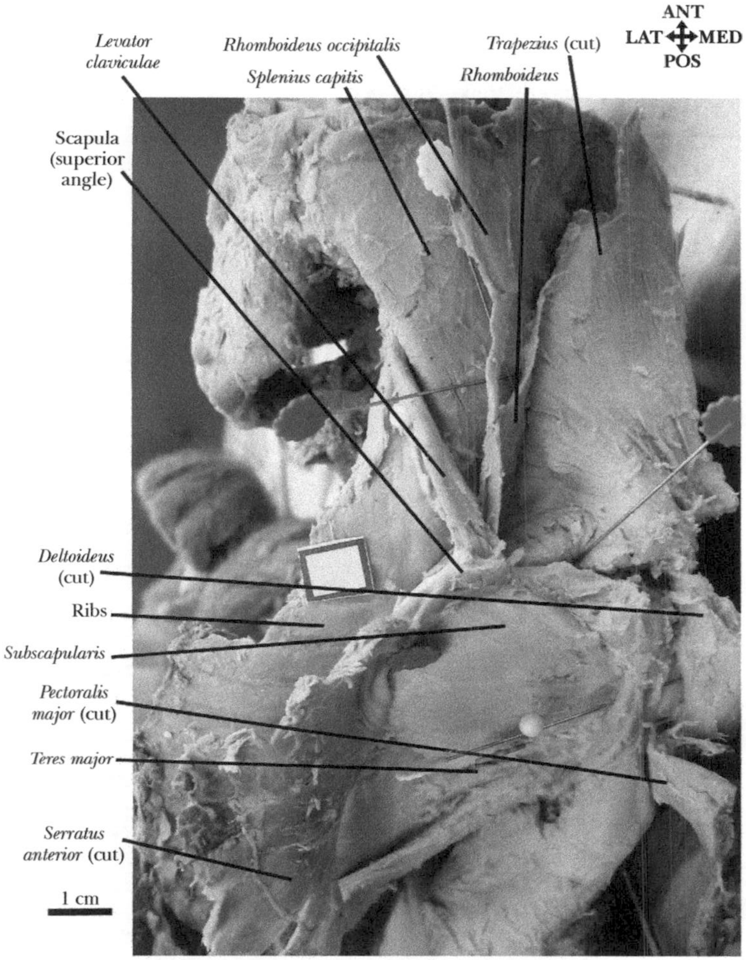

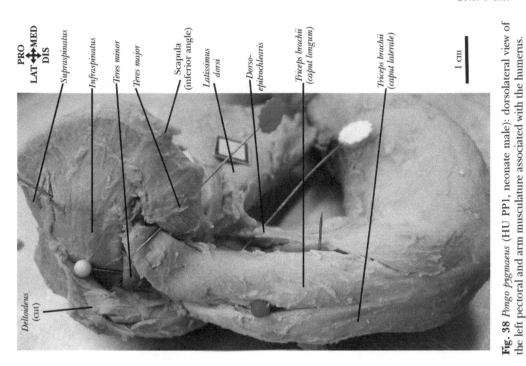

PRO
LAT ⬌ MED
DIS

Supraspinatus

Infraspinatus

Teres minor

Teres major

Scapula (inferior angle)

Latissimus dorsi

Dorso-epitrochlearis

Triceps brachii (caput longum)

Triceps brachii (caput laterale)

1 cm

Deltoideus (cut)

Fig. 38 *Pongo pygmaeus* (HU PP1, neonate male): dorsolateral view of the left pectoral and arm musculature associated with the humerus.

ANT
LAT ⬌ MED
POS

Trapezius (cut)

Rhomboideus

Rhomboideus occipitalis

Splenius capitis

Levator claviculae

Scapula (superior angle)

Deltoideus (cut)

Ribs

Subscapularis

Pectoralis major (cut)

Teres major

Serratus anterior (cut)

1 cm

Fig. 37 *Pongo pygmaeus* (HU PP1, neonate male): dorsolateral view of the left pectoral muscles after reflecting the trapezius and the pectoral girdle and its muscles in order to show the subscapularis on the ventral side of the scapula.

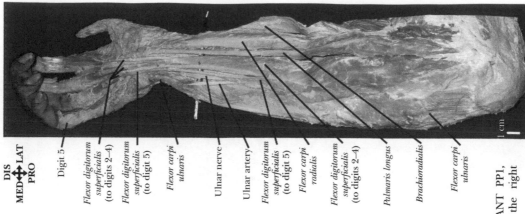

DIS
MED ← → LAT
PRO

Digit 5

Flexor digitorum superficialis (to digits 2–4)

Flexor digitorum superficialis (to digit 5)

Flexor carpi ulnaris

Ulnar nerve

Ulnar artery

Flexor digitorum superficialis (to digit 5)

Flexor carpi radialis

Flexor digitorum superficialis (to digits 2–4)

Palmaris longus

Brachioradialis

Flexor carpi ulnaris

Fig. 40 *Pongo pygmaeus* (GWUANT PP1, adult male): ventral view of the right forearm muscles.

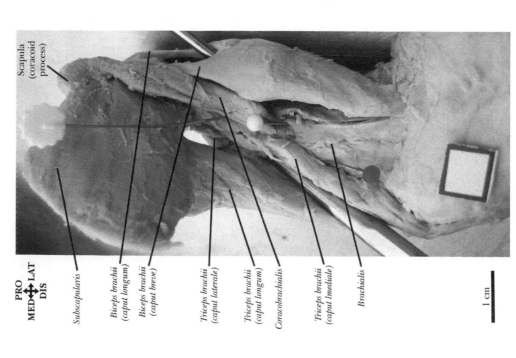

PRO
MED ← → LAT
DIS

Scapula (coracoid process)

Subscapularis

Biceps brachii (caput longum)

Biceps brachii (caput breve)

Triceps brachii (caput laterale)

Triceps brachii (caput longum)

Coracobrachialis

Triceps brachii (caput mediale)

Brachialis

1 cm

Fig. 39 *Pongo pygmaeus* (HU PP1, neonate male): ventrolateral view of the left pectoral and arm musculature associated with the humerus.

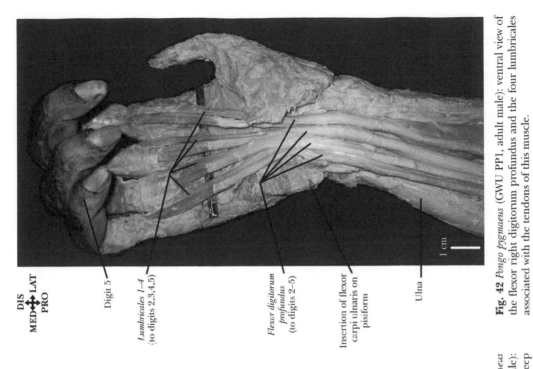

DIS
MED ⟷ LAT
PRO

Digit 5

Lumbricales 1–4
(to digits 2,3,4,5)

*Flexor digitorum
profundus*
(to digits 2–5)

Insertion of flexor
carpi ulnaris on
pisiform

Ulna

Fig. 42 *Pongo pygmaeus* (GWU PP1, adult male): ventral view of
the flexor right digitorum profundus and the four lumbricales
associated with the tendons of this muscle.

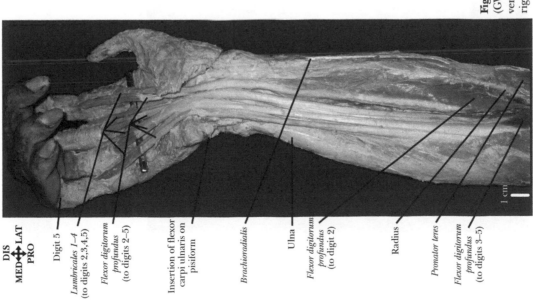

DIS
MED ⟷ LAT
PRO

Digit 5

Lumbricales 1–4
(to digits 2,3,4,5)

*Flexor digitorum
profundus*
(to digits 2–5)

Insertion of flexor
carpi ulnaris on
pisiform

Brachioradialis

Ulna

*Flexor digitorum
profundus*
(to digit 2)

Radius

Pronator teres

*Flexor digitorum
profundus*
(to digits 3–5)

Fig. 41 *Pongo pygmaeus*
(GWU PP1, adult male):
ventral view of the deep
right forearm muscles.

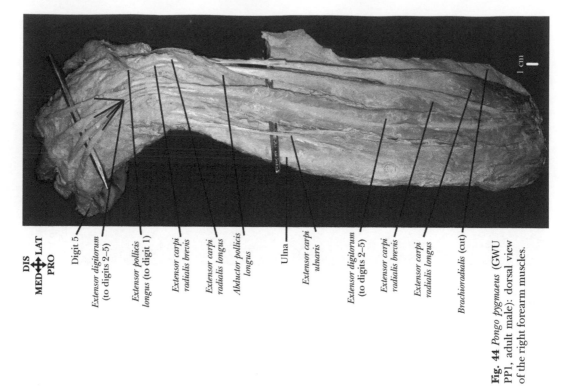

DIS
MED ⟷ LAT
PRO

Digit 5

Extensor digitorum
(to digits 2–5)

Extensor pollicis
longus (to digit 1)

Extensor carpi
radialis brevis

Extensor carpi
radialis longus

Abductor pollicis
longus

Ulna

Extensor carpi
ulnaris

Extensor digitorum
(to digits 2–5)

Extensor carpi
radialis brevis

Extensor carpi
radialis longus

Brachioradialis (cut)

1 cm

Fig. 44 *Pongo pygmaeus* (GWU PP1, adult male): dorsal view of the right forearm muscles.

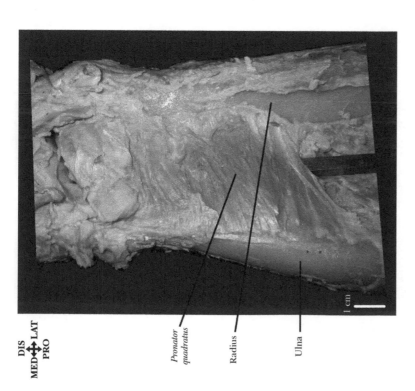

DIS
MED ⟷ LAT
PRO

Pronator
quadratus

Radius

Ulna

1 cm

Fig. 43 *Pongo pygmaeus* (GWU PP1, adult male): ventral view of the right pronator quadratus.

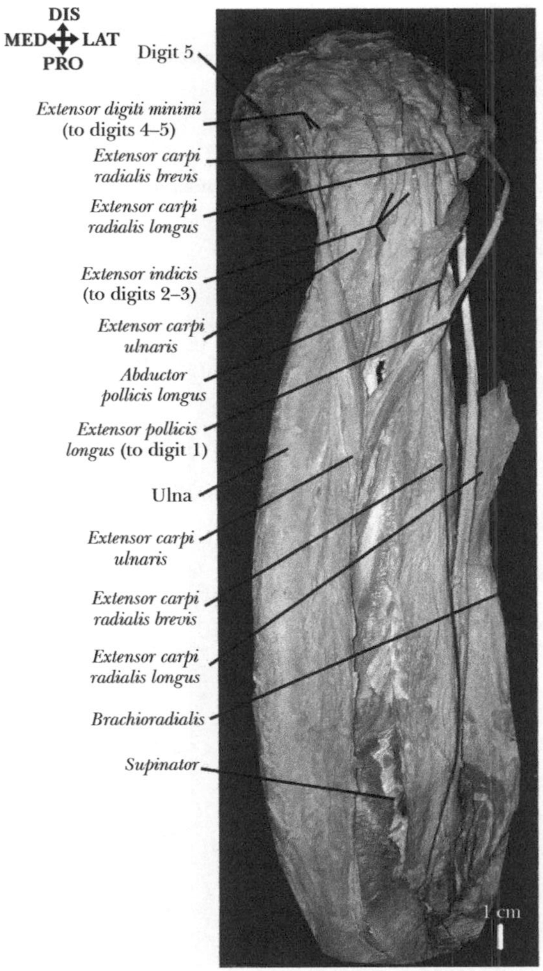

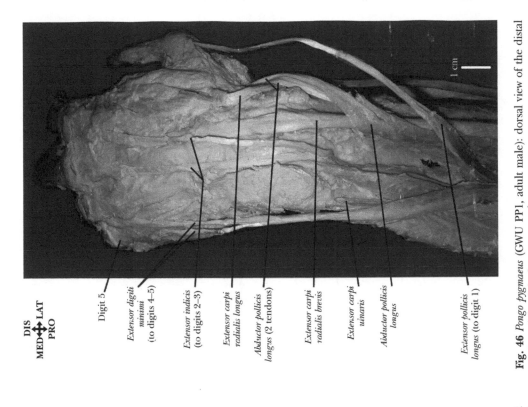

DIS
MED ⟵⊕⟶ LAT
PRO

Digit 5

*Extensor digiti
minimi
(to digits 4–5)*

*Extensor indicis
(to digits 2–3)*

*Extensor carpi
radialis longus*

*Abductor pollicis
longus (2 tendons)*

*Extensor carpi
radialis brevis*

*Extensor carpi
ulnaris*

*Abductor pollicis
longus*

*Extensor pollicis
longus (to digit 1)*

Fig. 46 *Pongo pygmaeus* (GWU PP1, adult male): dorsal view of the distal portion of the right forearm muscles.

DIS
MED ⟵⊕⟶ LAT
PRO

Digit 5

*Extensor digiti minimi
(to digits 4–5)*

*Extensor carpi
radialis brevis*

*Extensor carpi
radialis longus*

*Extensor indicis
(to digits 2–3)*

*Extensor carpi
ulnaris*

*Abductor
pollicis longus*

*Extensor pollicis
longus (to digit 1)*

Ulna

*Extensor carpi
ulnaris*

*Extensor carpi
radialis brevis*

*Extensor carpi
radialis longus*

Brachioradialis

Supinator

Fig. 45 *Pongo pygmaeus* (GWU PP1, adult male): dorsal view of the right forearm muscles; the extensor digitorum was removed, the extensor digiti minimi pulled medially and the extensor pollicis longus pulled laterally.

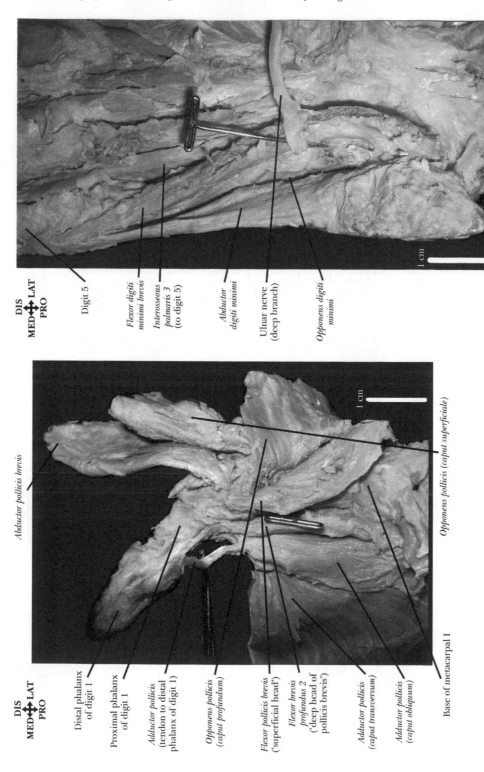

DIS
MED—LAT
PRO

Digit 5

Flexor digiti minimi brevis

Interosseous palmaris 3 (to digit 5)

Abductor digiti minimi

Ulnar nerve (deep branch)

Opponens digiti minimi

1 cm

Fig. 48 *Pongo pygmaeus* (GWU PP1, adult male): ventral (palmar) view of the right hypothenar muscles.

DIS
MED—LAT
PRO

Abductor pollicis brevis

1 cm

Opponens pollicis (caput superficiale)

Distal phalanx of digit 1

Proximal phalanx of digit 1

Adductor pollicis (tendon to distal phalanx of digit 1)

Opponens pollicis (caput profundum)

Flexor pollicis brevis ('superficial head')

Flexor brevis profundus 2 ('deep head of pollicis brevis')

Adductor pollicis (caput transversum)

Adductor pollicis (caput obliquum)

Base of metacarpal I

Fig. 47 *Pongo pygmaeus* (GWU PP1, adult male): ventral (palmar) view of the right thenar muscles.

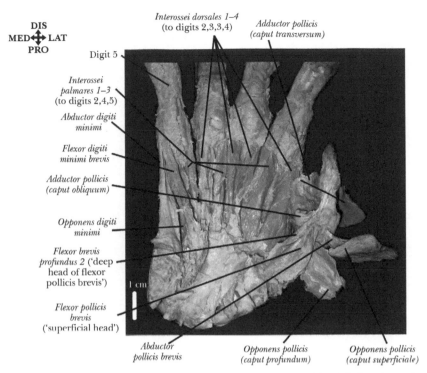

DIS
MED ✛ LAT
PRO

Interossei dorsales 1–4
(to digits 2,3,3,4)

Adductor pollicis
(caput transversum)

Digit 5

Interossei
palmares 1–3
(to digits 2,4,5)

Abductor digiti
minimi

Flexor digiti
minimi brevis

Adductor pollicis
(caput obliquum)

Opponens digiti
minimi

Flexor brevis
profundus 2 ('deep
head of flexor
pollicis brevis')

1 cm

Flexor pollicis
brevis
('superficial head')

Abductor
pollicis brevis

Opponens pollicis
(caput profundum)

Opponens pollicis
(caput superficiale)

Fig. 49 *Pongo pygmaeus* (GWU PP1, adult male): ventral (palmar) view of the right hand muscles after removal of the palmaris brevis and of the lumbricales.

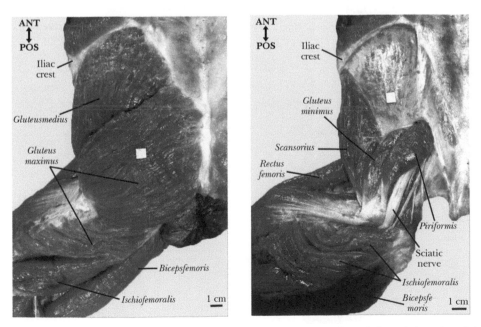

ANT
POS

Iliac
crest

Gluteusmedius

Gluteus
maximus

Bicepsfemoris

Ischiofemoralis

1 cm

ANT
POS

Iliac
crest

Gluteus
minimus

Scansorius

Rectus
femoris

Piriformis

Sciatic
nerve

Ischiofemoralis

Bicepsfe
moris

1 cm

Fig. 50 *Pongo pygmaeus* (VU PP1, adult female): dorsal view of the superficial (on left) and deep (on right) left buttock musculature.

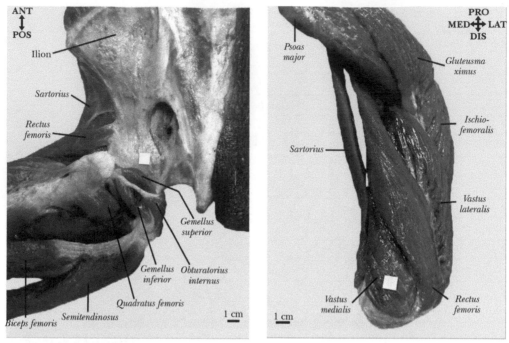

Fig. 51 *Pongo pygmaeus* (VU PP1, adult female): dorsal view of the deepest left buttock musculature (on left); ventral view of the superficial left thigh musculature (on right).

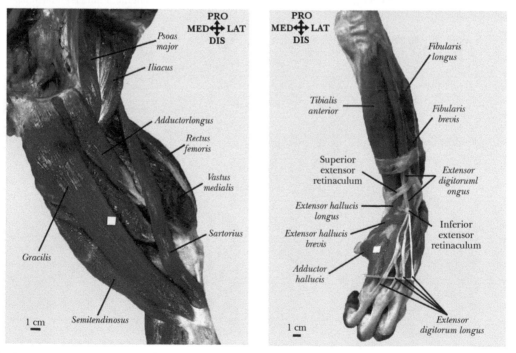

Fig. 52 *Pongo pygmaeus* (VU PP1, adult female): medial view of the superficial adductor musculature of the left thigh (on left); ventral view of the extensor musculature of the left leg (on right).

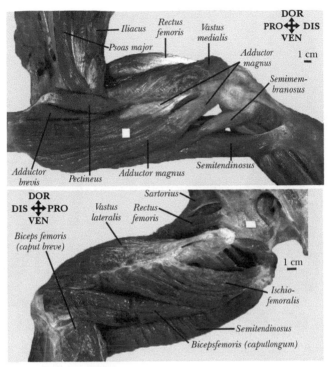

Fig. 53 *Pongo pygmaeus* (VU PP1, adult female): medial view of the deep adductor musculature of the left thigh (on top); lateral view of the superficial adductor musculature of the left thigh (on bottom).

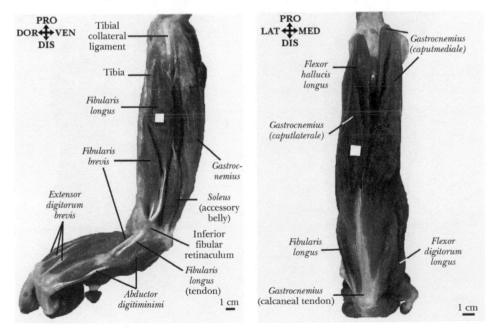

Fig. 54 *Pongo pygmaeus* (VU PP1, adult female): lateral view of the fibular musculature (on left) and dorsal view of the superficial flexor musculature (on right) of the left leg.

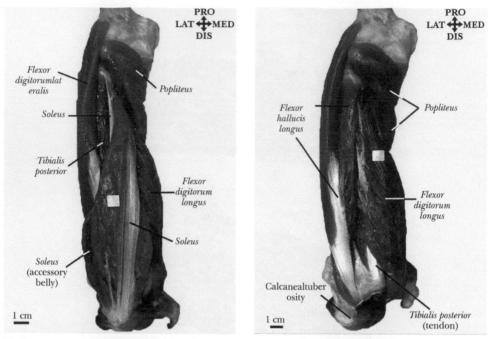

Fig. 55 *Pongo pygmaeus* (VU PP1, adult female): dorsal view of the deep (on left) and deepest (on right) flexor musculature of the left leg.

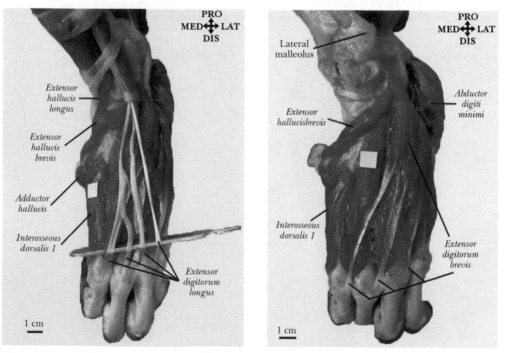

Fig. 56 *Pongo pygmaeus* (VU PP1, adult female): dorsal view of the superficial (on left) and deep (on right) extensor musculature of the left leg and foot.

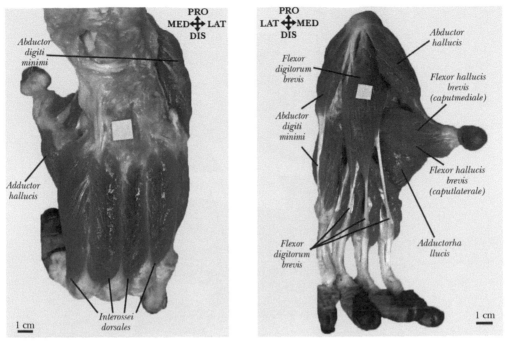

Fig. 57 *Pongo pygmaeus* (VU PP1, adult female): dorsal view of the deep musculature (on left) and ventral (plantar) view of the superficial musculature (on right) of the left foot.

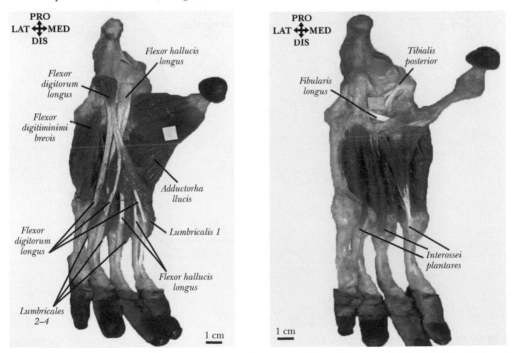

Fig. 58 *Pongo pygmaeus* (VU PP1, adult female): ventral (plantar) view of the superficial (on left) and deep (on right) flexor musculature of the left foot.

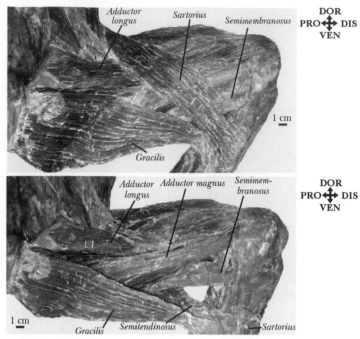

Fig. 59 *Pongo pygmaeus*(GWU PP1, adult male): medial view of the adductor musculature of the left thigh (on top; note the insertion of the sartorius covering that of the gracilis and of the semitendinosus).

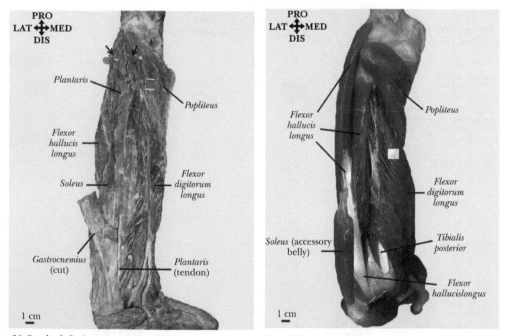

Fig. 60 On the left: *Pongo pygmaeus* (GWU PP1, adult male), dorsal view of the flexor musculature of the left leg; note the two heads of insertion (arrows) and the elongated muscular belly of the plantaris. On the right: *Pongo pygmaeus* (VU PP1, adult female), dorsal view of the flexor musculature of the left leg; note the accessory belly of the soleus, which is said to occur in approximately 3% of modern humans (accessory soleus *sensu* Moore & Dalley 2002).

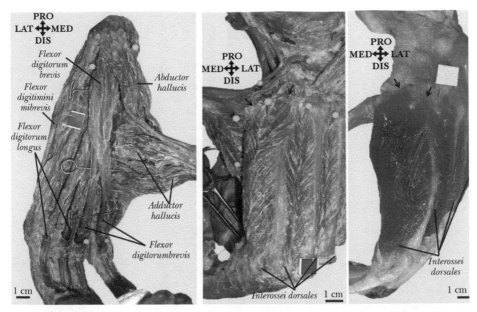

Fig. 61 On the left: *Pongo pygmaeus* (GWU PP1, adult male), ventral (plantar) view of the superficial flexor musculature of the left foot; note that the flexor digitorum brevis inserts onto digits 2 and 3 only. On the center: *Pongo pygmaeus* (GWU PP1, adult male), dorsal view of the interosseidorsales, note the two heads of insertion of the interosseusdorsalis 1 (from metatarsals 1 and 2; arrows). On the right: *Pongo pygmaeus* (VU PP1, adult female), dorsal view of the interosseidorsales; note the two heads of insertion of the interosseusdorsalis 1 (also from metatarsals 1 and 2; arrows).

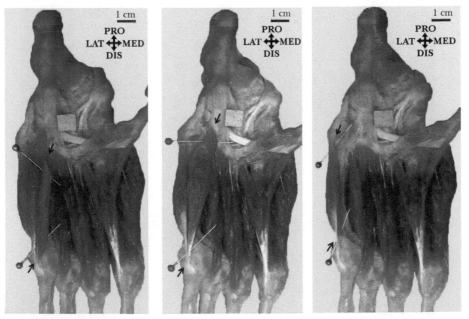

Fig. 62 *Pongo pygmaeus* (VU PP1, adult female): ventral plantar view of the deep musculature of the left foot. On the left, note the opponensdigitiminimi (arrows); on the center, note the presence of the flexor digitiminimibrevis (arrows); on the right, note the presence of the 'abductor os metatarsi digitiminimi' (arrows).

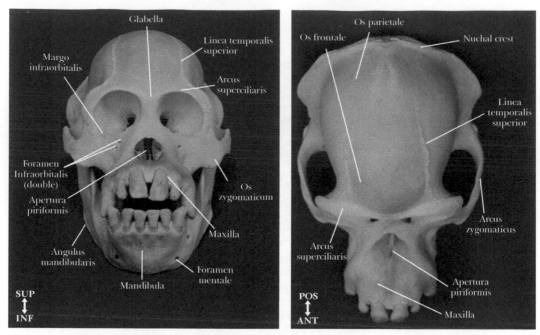

Fig. 63 *Pongo abelli* (VU PA1, adult female): facial (on left) and superior (on right) views of the cranium.

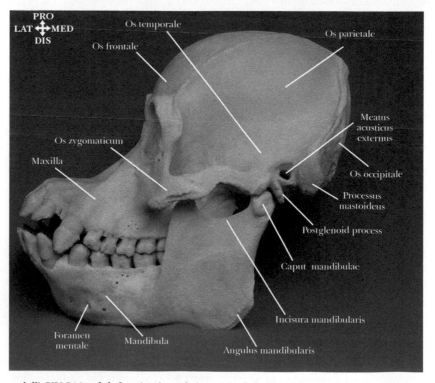

Fig. 64 *Pongo abelli* (VU PA1, adult female): lateral view of the left side of the cranium.

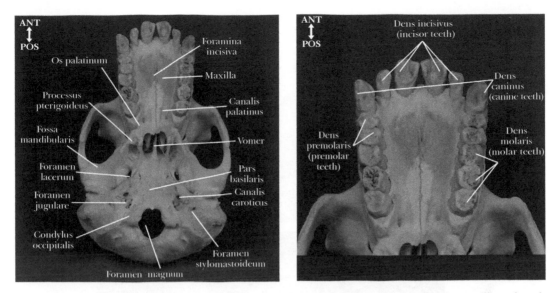

Fig. 65 *Pongo abelli* (VU PA1, adult female): inferior view of the cranium (on left); detail of the maxilla and teeth (on right).

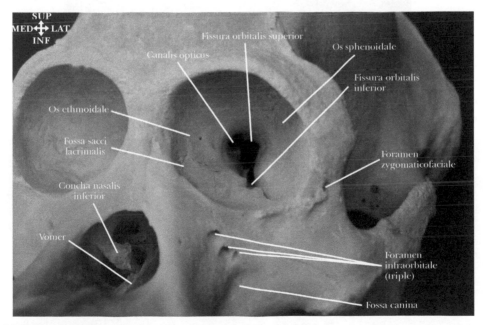

Fig. 66 *Pongo abelli* (VU PA1, adult female): detail of the left orbital cavity.

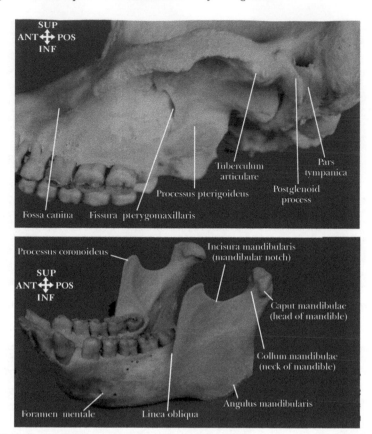

Fig. 67 *Pongo abelli* (VU PA1, adult female): details of the maxilla and the infratemporal fossa (on top); lateral view of the mandible (on bottom).

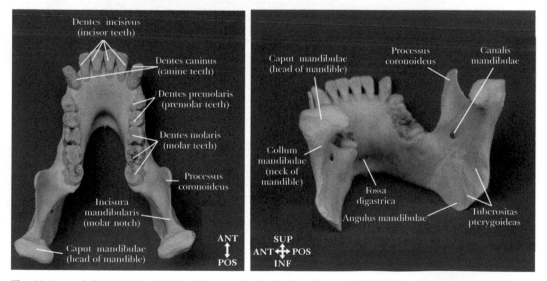

Fig. 68 *Pongo abelli* (VU PA1, adult female): superior (on left) and posterior (on right) views of the mandible.

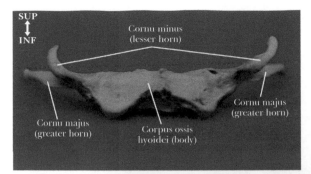

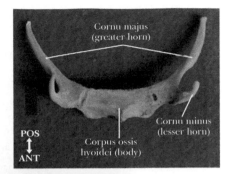

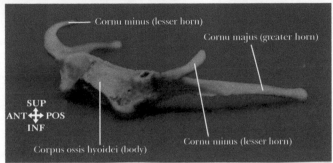

Fig. 69 *Pongo pygmaeus* (VU PP2, adult female): anterior (on top, left), superior (on top, right) and anterolateral (on bottom) views of the hyoid bone.

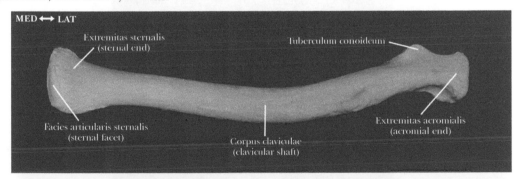

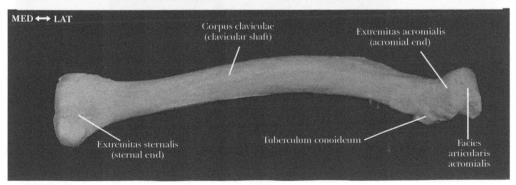

Fig. 70 *Pongo abelli* (VU PA1, adult female): anterior (on top) and posterior (on bottom) views of the left clavicle.

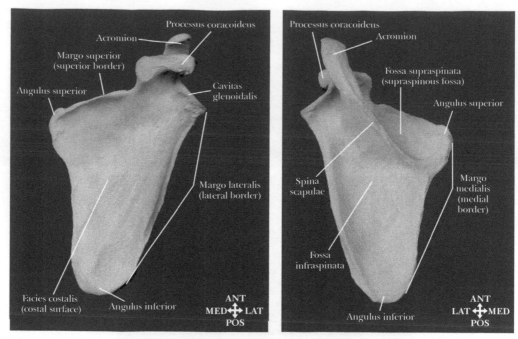

Fig. 71 *Pongo abelii* (VU PA1, adult female): ventral (on left) and dorsal (on right) views of the left scapula.

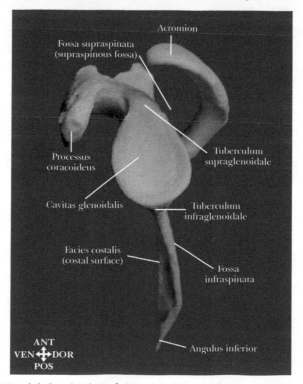

Fig. 72 *Pongo abelii* (VU PA1, adult female): lateral view of the left scapula.

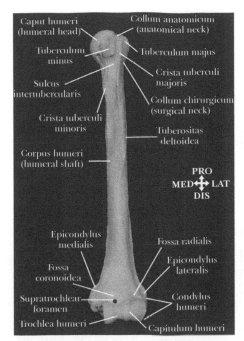
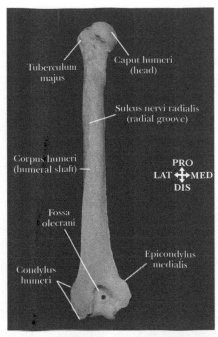

Fig. 73 *Pongo abelli* (VU PA1, adult female): ventral (on left) and dorsal (on right) views of the left humerus.

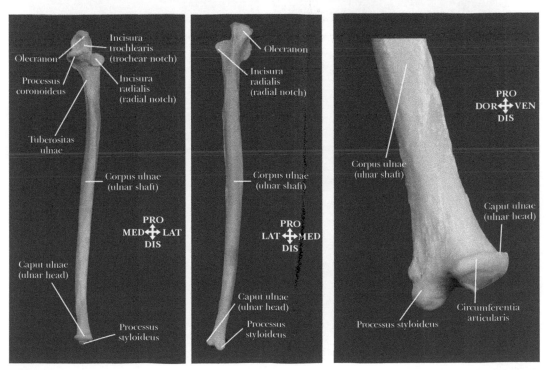

Fig. 74 *Pongo abelli* (VU PA1, adult female): ventral (on left) and dorsal (on center) views of the left ulna; details of the distal end of the left ulna (on right).

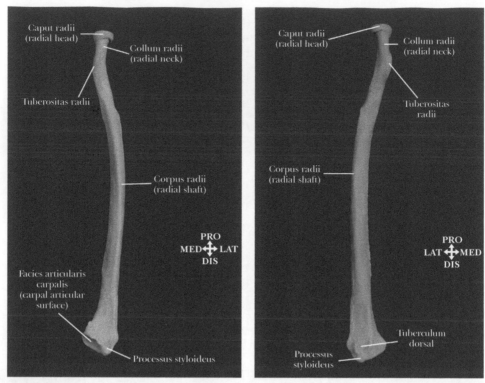

Fig. 75 *Pongo abelli* (VU PA1, adult female): ventral (on left) and dorsal (on right) views of the left radius.

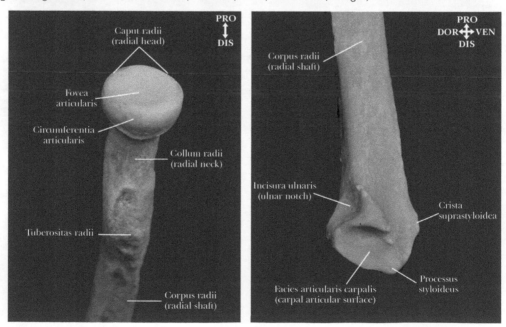

Fig. 76 *Pongo abelli* (VU PA1, adult female): medial view of the proximal (on left) and distal (on right) ends of the left radius.

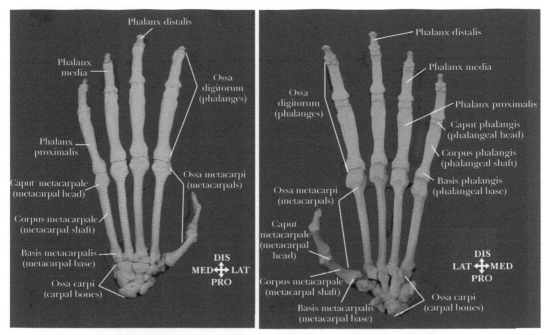

Fig. 77 *Pongo pygmaeus* (VU PP2, adult female): dorsal (on left) and ventral (palmar; on right) views of the left hand.

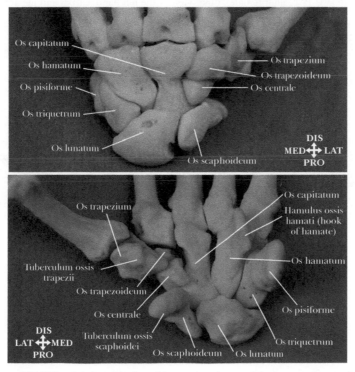

Fig. 78 *Pongo pygmaeus* (VU PP2, adult female): dorsal (on top) and ventral (palmar; on bottom) views of the carpal bones of the left hand.

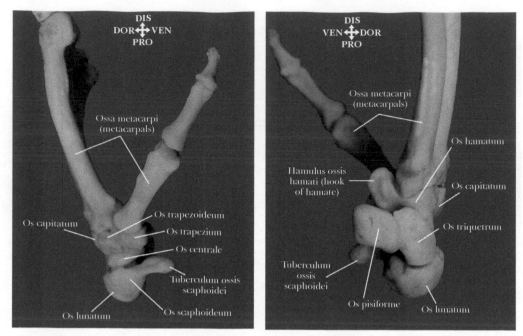

Fig. 79 On left: *Pongo pygmaeus* (VU PP2, adult female), radial view of the left hand (carpal and metacarpal bones). On right: *Pongo abelli* (VU PA2, adult female), ulnar view of the left hand (carpal and metacarpal bones).

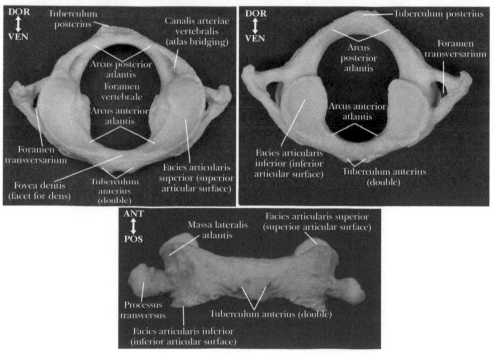

Fig. 80 *Pongo abelli* (VU PA1, adult female): anterior (top, left), posterior (top, right) and ventral (bottom) views of the atlas.

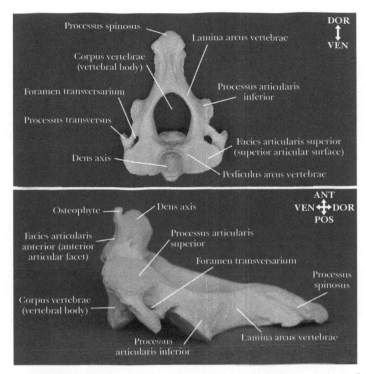

Fig. 81 *Pongo abelli* (VU PA1, adult female): anterior (on top) and lateral (on bottom) views of the axis.

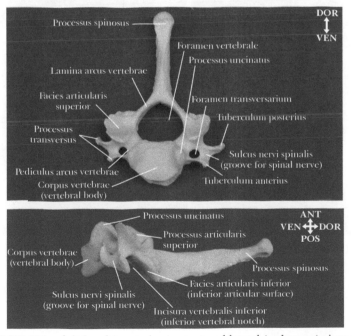

Fig. 82 *Pongo abelli* (VU PA1, adult female): anterior (on top) and lateral (on bottom) views of a typical cervical vertebra.

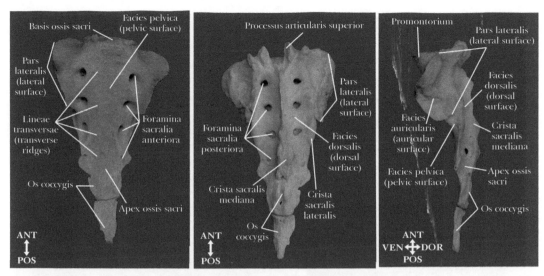

Fig. 85 *Pongo abelli* (VU PA1, adult female): ventral (on left), dorsal (on center) and lateral (on right) views of the sacrum.

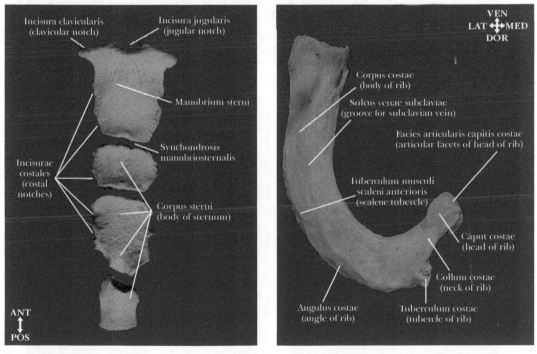

Fig. 86 On left: *Pongo pygmaeus* (VU PP2, adult female), ventral view of the sternum (xiphoid process is not ossified). On right: *Pongo abelli* (VU PA1, adult female), anterior view of the first left rib.

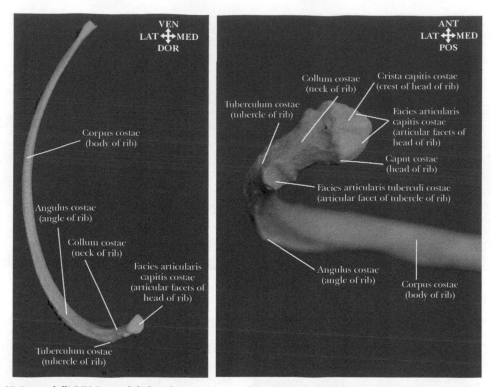

Fig. 87 *Pongo abelli* (VU PA1, adult female): anterior (on left) and dorsal (on right) views of the seventh left rib.

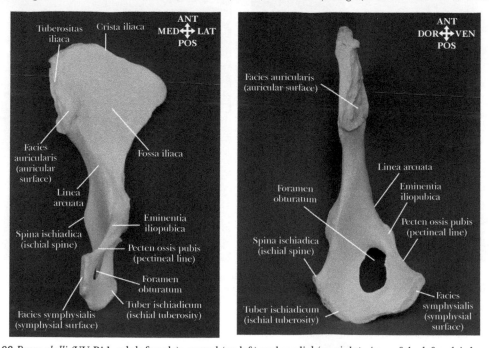

Fig. 88 *Pongo abelli* (VU PA1, adult female): ventral (on left) and medial (on right) views of the left pelvic bone.

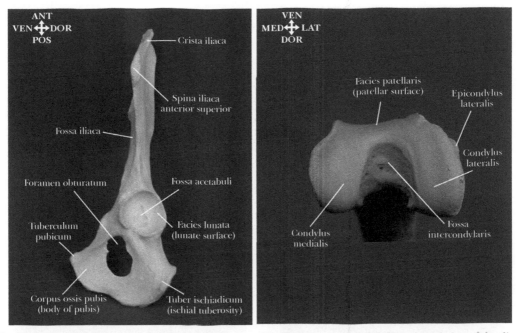

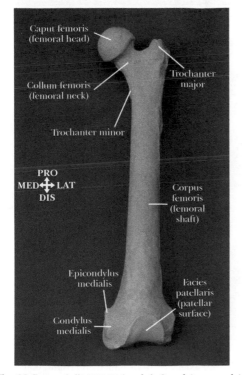

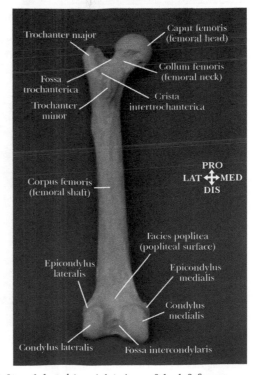

Fig. 89 *Pongo abelli* (VU PA1, adult female): lateral view of the left pelvic bone (on left) and axial view of the distal end of the left femur (on right).

Fig. 90 *Pongo abelli* (VU PA1, adult female): ventral (on left) and dorsal (on right) views of the left femur.

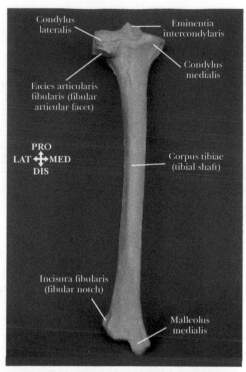

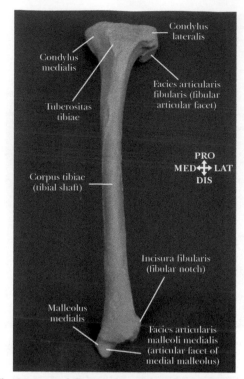

Fig. 91 *Pongo abelli* (VU PA1, adult female): ventral (on left) and dorsal (on right) views of the left tibia.

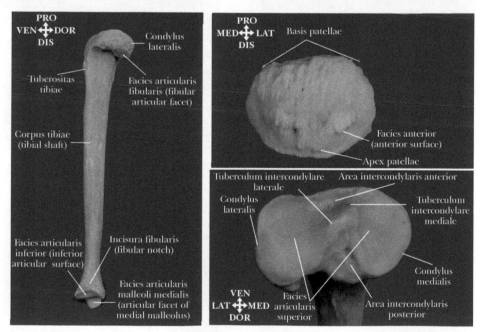

Fig. 92 *Pongo abelli* (VU PA1, adult female): lateral view (on left) and axial view of proximal end (right, bottom) of the left tibia; ventral view of left patella (right, top).

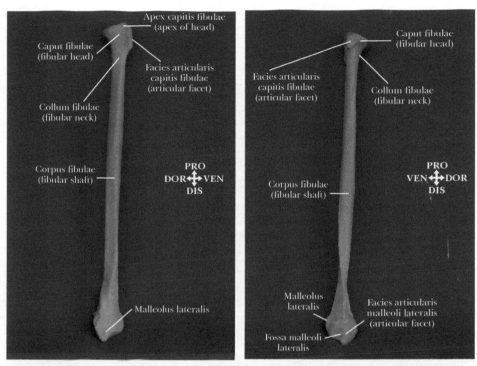

Fig. 93 *Pongo abelli* (VU PA1, adult female): lateral (on left) and medial (on right) views of the left fibula.

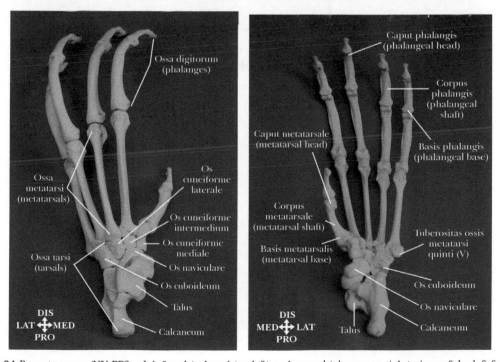

Fig. 94 *Pongo pygmaeus* (VU PP2, adult female): dorsal (on left) and ventral (plantar; on right) views of the left foot.

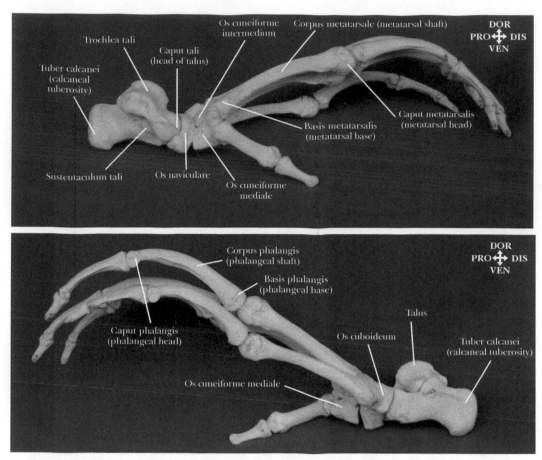

Fig. 95 *Pongo pygmaeus* (VU PP2, adult female): lateral (on bottom) and medial (on top) view of the left foot.

Milton Keynes UK
Ingram Content Group UK Ltd.
UKHW050451071024
449327UK00015B/326